Cathodic Protection of Steel in Concrete and Masonry

Second Edition

Cathodic Protection of Steel in Concrete and Masonry

Second Edition

Edited by
Paul M. Chess
John P. Broomfield

CRC Press
Taylor & Francis Group
Boca Raton   London   New York

CRC Press is an imprint of the
Taylor & Francis Group, an **informa** business

A SPON PRESS BOOK

CRC Press
Taylor & Francis Group
6000 Broken Sound Parkway NW, Suite 300
Boca Raton, FL 33487-2742

First issued in paperback 2017

© 2014 by Taylor & Francis Group, LLC
CRC Press is an imprint of Taylor & Francis Group, an Informa business

No claim to original U.S. Government works

Version Date: 20130830

ISBN 13: 978-1-138-07682-2 (pbk)
ISBN 13: 978-0-415-39503-8 (hbk)

Library of Congress Cataloging-in-Publication Data

Cathodic protection of steel in concrete
    Cathodic protection of steel in concrete and masonry / editors, Paul M. Chess, John P. Broomfield. -- Second edition.
        pages cm
    Revised edition of: Cathodic protection of steel in concrete. c1998.
    Includes bibliographical references and index.
    ISBN 978-0-415-39503-8 (hardcover : alk. paper)
    1. Reinforcing bars--Corrosion. 2. Reinforcing bars--Cathodic protection. 3. Reinforced concrete--Cathodic protection. I. Chess, Paul. II. Broomfield, John P. III. Title.

TA445.5.C38 2014
620.1'37--dc23   2013035164

Visit the Taylor & Francis Web site at
http://www.taylorandfrancis.com

and the CRC Press Web site at
http://www.crcpress.com

Contents

Editors

Dr. Paul M. Chess is specialised in reinforced concrete and building corrosion problems around the globe. He is the managing director of CP International, for which he has worked since 1994.

Dr. John Broomfield is an independent consulting engineer specialised in corrosion problems on large steel framed structures and corrosion of steel in concrete and has more than 25 years of experience in the design of cathodic protection systems for such structures.

Contributors

Chris Atkins is in Mott MacDonald, Altrincham, Cheshire, United Kingdom

John Broomfield is in Broomfield Consultants, Surrey, United Kingdom

Rene Brueckner is in Mott MacDonald, Altrincham, Cheshire, United Kingdom

Paul M. Chess is in CP International, Vallensbaek Strand, Denmark

Kevin Davies is in Corrociv Limited, Manchester, United Kingdom

Hernâni Esteves is in Ed. Züblin AG, Stuttgart, Germany

David Farrell is in Rowan Technologies, Manchester, United Kingdom

Tony Gerrard is in BAC Corrosion Control Limited, Shropshire, United Kingdom

Frits Gronvold is in Gronvold & Karnov, Vallensbaek Strand, Denmark

Ulrich Hammer is in Ed. Züblin AG, Stuttgart, Germany

Paul Lambert is in Mott MacDonald, Altrincham, Cheshire, United Kingdom

Arnoud Meillier is in ACS Associés, Gardanne, France

Richard Palmer is in Palmer Consulting, Divonne-les-Bains, France

Chapter 1

Corrosion in reinforced concrete structures

Paul M. Chess

CONTENTS

1.1 INTRODUCTION

Since very early times humans have used masonry structures, and for thousands of years they have secured stones or other parts of structures that might be suffering from tensile stresses with metal fixings. This type of structure has evolved into steel-framed buildings. An associated composite structural material of concrete reinforced with steel has risen rapidly to dominance. Steel and concrete have become the most common materials for man-made structures over the last 100 or so years with the use of the composite material, concrete reinforced with steel becoming one of the most popular methods for civil construction. The historical reasons for steel-reinforced concrete's popularity are not hard to find with its cheapness, high structural strength, mouldability, fire resistance and supposed imperviousness to the external environment while requiring little or no maintenance, providing a virtually unbeatable combination. To harness these properties, both national and international standards have been developed. The standards for both concrete and steel were initially principally defined by compositional limits and strength, and this has continued to be the primary means of quality control to date. The use of steel and metals in masonry buildings peaked in the 1920s and is still common today.

Up until the 1950s, it was assumed that when steel was encased in an alkaline concrete matrix neither would suffer any degradation for the indefinite future. However, evidence of degradation was noted as early as 1907[1] when it was observed that chlorides added to concrete could allow sufficient corrosion of the steel to crack the concrete. Many reinforced concrete structures have now reached their design lives without any evidence of structural degradation. However, it is now evident that in areas where there is an aggressive atmosphere the concrete can be damaged or the steel can corrode in a dramatically shorter period than that specified as their design life. For U.K. highways, the nominal design life is 120 years. However, it has been noted that highway structures are showing significant corrosion problems after a much shorter period than this. In extreme cases,[2] the estimated time to corrosion activation of steel reinforcement in modern concrete with the designated cover can be as low as 5.5 years at the 0.4% chloride level with modern concrete. These research findings are in accordance with site investigations. A substantial number of structures have been found to have their steel reinforcement sufficiently corroded within 20 years of construction to be structurally unsound.

The traditional use of cathodic protection has been to prevent the corrosion of steel objects in ground or water, and this is still its most common application. It is now almost universally adopted on ships, oil rigs, and oil and gas pipelines. Over the last 50 years, cathodic protection has advanced from being a black art to somewhere approaching a science for these applications.

Over the past 30 years or so, there has been a steady increase in the use of cathodic protection for the rehabilitation of reinforced concrete structures that are exhibiting signs of distress, and more recently it has been used to protect iron and steel in masonry structures. The most common damage mechanism for concrete structures is chloride-induced corrosion of steel reinforcement, and this is normally what cathodic protection systems are intended to stop. On masonry buildings, it is generally corrosion caused by the loss of the inhibiting effects of the surrounding mortar. Initially, cathodic protection techniques for reinforced concrete followed the practice of traditional impressed current systems closely; but particularly over the past decade or so, there have been significant developments that have allowed the protection of these structures to become a legitimate and yet distinctly different part of the cathodic protection mainstream with its own protection criteria, anode types and even power supplies. The protection of masonry buildings has followed a similar path, initially closely aping reinforced concrete designs but over time becoming more differentiated.

The objective of this book is to introduce the current state of the art in cathodic protection of steel in concrete and an allied, but separate, field of metals in masonry structures. Various aspects of the topic are introduced in the coming chapters to introduce the various subjects. The objectives are that a practising civil engineer, architect or owner should have an introduction into the multi-disciplined world of cathodic protection for a civil structure.

1.2 ELECTROCHEMICAL CORROSION

Electrochemical reactions are widely used by humankind for industrial processes such as anodising, or the production of chloride, and indeed are used directly by most people every day of their lives, for example, when using a battery. A surprising number of engineers vaguely remember an explanation in chemistry classes of how a battery operates. This is normally reiterated as being about electrolytes with ions swimming about, with anodes and cathodes making an appearance, and then dismissed as not being of importance in 'proper' civil or mechanical engineering. Unfortunately for those who do not like electrical circuits, corrosion is also an electrochemical process and is of great economic importance, as people with old cars will testify. Corrosion has been estimated to consume 4% of the gross national product of, for example, the United States.[3] This percentage is likely to be of the same order globally. The corrosion process is often the life-determining factor in many reinforced concrete or masonry structures, albeit the timescales to first apparent distress are normally very different.

In corrosion under normal atmospheric conditions, and all the aforementioned processes, an electrochemical cell is needed for the reactions to occur. This cell comprises an anode and a cathode separated by an electrolytic conductor with a metallic connection. This is shown schematically in Figure 1.1. A practical definition of an anode is the area where corrosion occurs, whereas a cathode is the area where no corrosion occurs.

When a metal such as steel is in an electrolyte (this is a water-based solution that has conductive ions such as sodium chloride in solution), a corrosion cell can be formed. Some of the steel in the electrolyte forms the anode (A in Figure 1.2), and a part of the steel that is also in the same electrolyte forms the cathode (C in Figure 1.2). Corrosion in this case would be occurring at all the anode points that are dispersed around the steel. This gives the appearance of general or uniform corrosion. In this case, the corroding metal is acting as a mixed electrode.

At anodic sites, metal atoms pass into the solution as positively charged ions (anodic oxidation) and the excess of electrons flow through the metal to cathodic sites where an electron acceptor like dissolved oxygen is available to consume them (cathodic reduction). This process is completed by

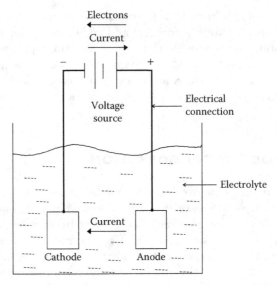

Figure 1.1 Schematic of a corrosion cell: in a driven cell, cations migrate towards the cathode and anions towards the anode. Current is defined as the flow of positive charge and moves in a direction opposite to the flow of electrons.

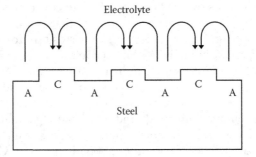

Figure 1.2 Schematic of micro-corrosion cells on steel surface: regions labelled A are the anodic areas where metal is dissolving. Regions labelled C are cathodic areas where no corrosion is occurring. The arrows represent the current flow.

the transport of ions through the aqueous electrolyte to produce soluble or insoluble corrosion products.

The corrosion reactions are

$$Fe = Fe^{2+} + 2e^- \quad \text{anodic corrosion reaction} \qquad (1.1)$$

$$O_2 + 2H_2O + 4e^- = 4OH^- \quad \text{cathodic reaction} \qquad (1.2)$$

$$2H_2O + 2e^- = H_2 + 2OH^- \quad \text{cathodic reaction} \qquad (1.3)$$

In alkaline and oxygen-rich electrolytes such as an atmospherically exposed reinforced concrete structure, either reaction II or reaction III or both can occur. The iron ions dissolved in the pore water of the concrete pass through several more stages to get to the final corrosion product. When chloride is present several complexes are formed, which reduce the activation energy, and this massively increases the rate of reaction:

$$Fe^{2+} + 2OH^- = Fe(OH)_2 \quad \text{ferrous hydroxide} \tag{1.4}$$

$$4Fe(OH)_2 + O_2 + 2H_2O = 4Fe(OH)_3 \quad \text{ferric hydroxide} \tag{1.5}$$

$$2Fe(OH)_3 = Fe_2O_3 \cdot H_2O + 2H_2O \quad \text{hydrated ferric oxide} \tag{1.6}$$

1.3 CORROSION OF STEEL

Steel in common with all engineering materials is intrinsically unstable in that it wants to return to its stable state where it came from as an ore. The result of this reversion is rust (commonly iron oxide, but can also be iron sulphide or other compounds), which while having a considerably greater chemical stability also has greatly reduced mechanical properties such as strength compared with the original steel. As there is this tendency to corrode, the principal question is not will steel rust but how fast will it rust?

The corrosion rate of steel is normally decided by the environment and also the stability of the oxide layer on its surface. If this layer forms a protective skin that is not breached, then the rate of reaction is very slow. If, however, the oxide layer is open at many places and sloughs off the surface, providing access for more oxygen (which is normally dissolved in water) to the unreacted steel surface, then a high corrosion rate can be expected.

Straight carbon and high-yield steels are the most commonly used grades for rebar in normal civil engineering projects. Neither of these types has a particularly protective oxide film, and both rely on the alkalinity of the concrete to stabilise this skin. This surface skin is a very dense oxide layer of the order of 5 nm. It has been conjectured that this film is a crystalline layer of Fe_3O_4 with an outer layer of γ-Fe_2O_3. More recently, it has been proposed that the structure is amorphous and polymeric in nature.

When a metal such as steel is in an electrolyte (this is a water based solution which has conductive ions such as sodium chloride in solution) then a corrosion cell can be formed. Some of the steel in the electrolyte forms the anode, and part of the steel which is also in the same electrolyte forms the cathode. Corrosion in this case would be occurring at all the anode points which are dispersed around the steel. This gives the appearance of general or uniform corrosion. In this case, the corroding metal is acting as a mixed electrode.

At anodic sites, metal atoms pass into solution as positively charged ions (anodic oxidation) and the excess of electrons flow through the metal to cathodic sites where an electron acceptor like dissolved oxygen is available to consume them (cathodic reduction). This process is completed by the transport of ions through the water based electrolyte to produce soluble or insoluble corrosion products.

When steel corrodes in a normal atmosphere, that is, outdoors, there will be a rapid change in colour. This is known as 'flash' rusting. As an example, blast-cleaned steel in a moist environment changes colour in the time between the contractor finishing the blasting operation and opening the paint pots. This rusting is evidenced by a change in the surface colour from silver to orange–red over the entire exposed surface. In this case, the corrosion is very rapid because of the presence of ample fuel (oxygen) and the absence of a protective oxide film. In a saline environment flash rusting is even quicker as chloride helps the water to conduct current, thus completing the corrosion reaction's electrical circuit. If the steel were examined visually under a microscope it would all look uniform, as in this case the individual anode and cathode sites are very small, perhaps within a few microns of each other.

In cases where steel is exposed directly to the atmosphere, it is at a normal (neutral) pH and there is a reasonable supply of oxygen, there will be widespread and uniform corrosion. This is normally observed when large sections of steel are rusting and can be seen on any uncovered steel article particularly on beaches and other places with an aggressive atmosphere.

When the access of oxygen to steel is reduced and this becomes the corrosion-limiting step, that is, when there are sufficient aggressive ions present at the steel interface so that the corrosion reaction itself can happen very quickly, then other forms of corrosion may occur. The most common is pitting corrosion. This, for example, will occur when there is a surface coating on the steel that is breached, allowing oxygen and moisture access to a relatively small area. In older cars, these are commonly seen as rust spots. This situation is shown schematically in Figure 1.3.

1.4 STEEL IN CONCRETE

Concrete normally provides embedded steel with a high degree of protection against corrosion. One reason for this is that cement, which is a constituent of concrete, is strongly alkaline. This means that the concrete surrounding the steel provides an alkaline environment for the steel. This stabilises the oxide or hydroxide film and thus reduces the oxidation rate (corrosion rate) of the steel. This state with a very low corrosion rate is termed passivation. The other reason why concrete provides embedded steel with protection is that it provides a barrier to outside elements that are aggressive to the steel. The most common agent for depassivation of steel in concrete is the chloride ion.

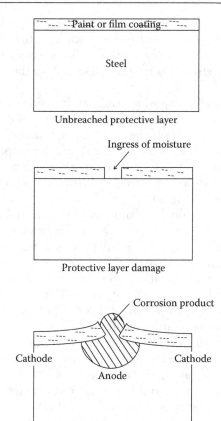

Figure 1.3 Example of pitting corrosion.

In the 1950s and 1960s, it was assumed that concrete of low water–cement ratio, which was well cured, would have a sufficiently low permeability to prevent significant penetration of corrosion-inducing factors such as oxygen, chloride ions, carbon dioxide and water. Unfortunately, this has not been found to be the case. Some of deviation from the predicted behavior can be explained by the fact that concrete is inherently porous, whatever its composition, and if there is a concentration gradient then at some time a sufficient quantity of aggressive ions will be passed through the concrete to initiate corrosion. In most structures there are many cracks, and these can provide preferential pathways for corrosion-inducing factors. The crux of the time for initiation is that this 'some time' is sufficiently long to achieve the design life. In the past 20 years, significant programs have been undertaken by various authorities to model the rates of chloride ingress and verify these models with site data.

Fortunately, in the majority of steel-reinforced concrete structures corrosion does not occur in the design life. In principle, with concrete of suitable quality, corrosion of steel can be prevented for a certain period, provided that the structure or element is properly designed for the intended environmental exposure.

In instances of severe exposure, such as in bridge decks exposed to de-icing salts or piles in flowing sea water, the permeability of concrete is of critical importance to the life of the structure and further protection methods other than the application of concrete should be adopted (Figure 1.4).

In electrochemical corrosion, the flow of electrical current and one or more chemical processes are required for there to be metal loss. The flow of electrical current can be caused by 'stray' electrical sources such as from a train traction system or from large differences in potential between parts of a structure caused by factors such as differential aeration from the movement of sea water (the mechanism for this is still uncertain, but it could be that very large cathode areas are built up in the tidal zone because of oxygen charging). The incidence of electrochemical corrosion by these electrical current sources alone is rare but can be serious when it occurs. Often, this process can contribute to the corrosion rate when there are other aggressive factors.

It is likely that the passivation on steel by alkalinity would allow a certain amount of current discharge from the steel without metal loss. The critical factor in this is the resupply of alkalinity relative to the current drain. An example of stray current corrosion is a jetty where the piles were being cathodically protected and the reinforced concrete deck was being used as the system negative. Unfortunately, several of the piles were electrically discontinuous and corrosion occurred at a secondary anode point

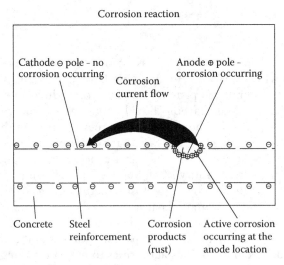

Figure 1.4 Corrosion cell in steel reinforcement.

formed on these piles as the current attempted to flow back to the system negative. In this case the large amount of current flowing means that the corrosion was severe.

The vast majority of potential gradients found between different areas of steel in reinforced concrete are caused by the existence of physical differences or non-uniformities on the surface of the steel reinforcement (different steels, welds, active sites on the steel surface, oxygen availability and chloride contamination). These potential gradients can allow significant electrical current to flow and cause under certain circumstances, that is, with aggressive ions in the concrete, severe corrosion of the reinforcement.

Even though the potential for electrochemical corrosion might exist because of the non-uniformity of the steel in the concrete, this corrosion is normally prevented even at nominally anodic sites (i.e. more negative in potential than cathodic areas) by the passivated film that is found on the steel surface in the presence of moisture, oxygen and water-soluble alkaline products formed during the hydration of the cement.

There are two mechanisms by which the highly alkaline environment and accompanying passivation effect may be destroyed, namely the reduction of alkalinity by the leaching of alkaline substances by water or neutralisation when reacting with carbon dioxide or other acidic materials. A second mechanism is by electrochemical action involving aggressive ions acting as catalysts (typically chloride) in the presence of oxygen.

Reduction of alkalinity by reaction with carbon dioxide, as present either in air or dissolved in water, involves neutralising reactions with sodium and potassium hydroxides and subsequently the calcium system, which are part of the concrete matrix. This process called carbonation, although progressing increasingly slowly, may in time penetrate the concrete to a depth of 25 mm or so (depending on the quality of the concrete and other factors) and thereby neutralise the protective alkalinity normally afforded to steel reinforcement buried to a lesser depth than this. This form of damage is particularly apparent in low-grade concrete structures where builders were economical with the cement and liberal with the water.

The second mechanism where the passivity of steel in concrete can be disrupted is by electrochemical action involving chloride ions and oxygen. As previously mentioned, this is by far the most important degradation mechanism for reinforced concrete structures, and the most significant factors influencing this reaction are discussed in Section 1.4.1.

1.4.1 Alkalinity and chloride concentrations

The high alkalinity of the chemical environment normally present in concrete protects the embedded steel because of the formation of a protective film, which could be either an oxide or a hydroxide or even something in the middle depending on which research paper you read. The integrity and protective quality of this film depend on the alkalinity (pH) of the environment. The bulk alkalinity of the concrete depends on the water-soluble alkaline products. The principal soluble product is calcium hydroxide, and the initial alkalinity of the concrete is at least that of saturated lime water (pH of about 12.4 depending on the temperature). In addition, there are relatively small amounts of sodium and potassium oxides in the cement, which can further increase the alkalinity of the concrete or paste extracts, and pH values of 13.2 and higher have been reported.

The higher the alkalinity, the greater the protective quality of the protective film. Steel in concrete becomes potentially more susceptible to corrosion as the alkalinity is reduced. Also, steel in concrete becomes more at risk with increasing quantities of soluble chlorides present at the iron–cement paste interface. Chloride ions appear to be a specific destroyer of the protective oxide film.

As chloride ion levels increase within the concrete adjacent to the steel, two competing mechanisms fight for dominance on the steel surface. These are stabilisation and repair of the oxide film on the surface of the steel by hydroxyl ions and the disruption of the film with a reaction between the steel metal and the chloride ions.

The form of the oxide layer on the surface of the steel has been discussed previously, but where breached there is a rapid reaction (III) at a rate on the order of 1 μs, between the steel and the chloride. This leaves a microscopic patch of metal chloride at the steel to oxide film interface, and this is probably the initiator of the corrosion pit. The high speed of this reaction would suggest that this is not the rate-limiting step in the corrosion process, and other mechanisms determine the speed at which the reaction advances.

It has been widely accepted that there will be an initiation of corrosion on the steel at a certain chloride to hydroxyl concentration; however, quantitatively the results show a significant disparity even in alkaline solutions with Hausman[4] arriving at a chloride ion concentration 0.6 times that of the hydroxyl concentration. Gouda[5] reported a ratio of 0.3, although this was in sodium hydroxide and not calcium hydroxide. Neither of these experiments was in real concrete (both were in alkaline solutions) and neither imposed control over the oxygen level.

The action of the chloride ions has been reported variously in three ways[6]:

1. Chloride ions pass directly through the amorphous oxide film.
2. Chemical adsorption of chlorides on to the steel surface.
3. Chlorides compete with hydroxyl ions for the ferrous ions in steel. This complex of ferrous chloride then breaks away from the steel surface, developing the passive layer breakdown.

The first stage of the metal and chloride reaction is 'green rust'. This is variously characterised as a complex of intermediate compounds, which in the presence of sufficient oxygen will allow the iron ions to reach its more stable trivalent state where the oxide does not have the chloride compounded with it. This ejection of chloride acts as a concentration mechanism, and thus it is likely that after initiation a lower overall chloride concentration level will then be sufficient to maintain the corrosion rate. This is significant as in real structures the chloride ions will generally reach the rebar at higher concentrations in discrete locations and can explain why there is often a morphological difference in the corrosion structure observed after breakout in oxygenated and non-oxygenated parts of structures. It is found that in low-oxygen areas the pits tend to be deep and sporadic, whereas in high-oxygen availability areas there is more widespread and less deep corrosion.

In actual concrete highway structures, Vassie studied corrosion incidence, measured by physical examination of the steel after breakout, against chloride level and found isolated corrosion areas down to very low chloride levels with the incidence increasing as the chloride level increased. This was followed by Podler[7] who found a similar incidence. What is significant is that neither set of data appears to suggest that there is a definite corrosion initiation level.

However, the aforementioned data was described by Bentur et al.[8] as 'hardly any corrosion occurs below 0.4% chloride content; increase in corrosion rate starts at levels above 1% chlorides by weight of cement'. This finding was backed by an earlier work[8] in which calcium chloride introduced into the mix showed that the corrosion rate was low below 2.3% and thereafter increased in a linear fashion as the chloride level increased. Again, a definite corrosion initiation level was not found.

One of the factors that will vary the amount of chloride required to initiate corrosion significantly is the amount of chloride that is bound in the concrete; typically, this will be in the form of the relatively insoluble tricalcium chloroaluminate, commonly known as Friedel's salt. This bound chloride is held in the concrete structure, and it is only the 'free' chloride in the pore water that is available to take part in the corrosion reaction. The amount of binding of the chloride is normally considered to be a function of the tricalcium aluminate (C_3A) level of concrete.

Another possible factor affecting the initiation level is the formation of a dense trivalent oxide coating (mill scale) through the hot rolling and quenching of the rebar during manufacture.

A further factor that will have a significant effect on the initiation level is electrochemical potential. If the steel was isolated, this potential would be set by the reactivity of the steel at its electronic to ionic interface, that is, at the steel to oxide dividing line, at the one particular location. In a typical structure, much, or all, of the steel is in some sort of electronic contact and potential differences are likely to exist throughout the structure with consequent current flows. It is likely that the passive oxide film on the steel surface would limit these flows while the metal chlorides would aid them (chlorides have a similar ionic conductance to most other cations but have water molecules associated with the compound, which allow hydroxides to move. These have about three times the ionic mobility of other cations). These current flows could assist the chloride ions' penetration of the oxide barrier layer and thus substantially reduce the initiation level. The potential change will also change the movements of the ions particularly in the double layer, which constitutes the reaction zone.

Chloride can be present in 'as manufactured' concrete as a set accelerator (calcium chloride) or can enter through contamination of the concrete mix, but more commonly the chloride comes from an external source such as de-icing salts or marine environments. In these latter cases, the salt diffuses through the concrete cover to the steel. It is worth noting that in practice the rate of movement of chloride may be very different from the 'chloride diffusion coefficient' that is used to assess the durability of the structure. This is because concentration diffusion is only one transport mechanism. Transport of chlorides can also utilise convection flow, capillary suction and electro-osmosis. The movement of chlorides may be restricted by chloride binding or interaction. In real structures, cracking and the mechanical movement of salt water through these opening and closing defects needs careful consideration, especially in view of recent developments of high-strength concrete mixes with low permeability and low ductility.

It is sometimes found that reinforced concrete with uniformly high levels of chloride contamination (often over 3%) does not have significant corrosion of the rebar. This typically would happen where there are constant environmental conditions around the concrete, for example, internal walls of a building or structures buried below a saline water table. Conversely, areas where there are cyclical environmental conditions, such as where concrete is exposed to strong tidal flows of aerated salt water or where there are diurnal weather conditions, for example, where there is direct sunlight in the day and high humidity and low temperatures in the night, there can be very significant damage at low chloride concentrations and a young age.

Various ideas on what the chloride is doing to cause this depassivation have been proposed, but there seems to be agreement that in localised areas the passive film is broken down, resulting in pitting. In the pits an acid environment exists, and when concrete is stripped from the corrosion sites on steel green–black and yellow–black compounds can often be observed. These are probably intermediate complexes that contain chloride and allow a lower activation energy for the oxidation process. In the corrosion process, the chloride is not normally held as a final product and can be thought of as acting as a catalyst.

Any increase in chloride ion concentration beyond the initiation level is likely to increase the rate of corrosion. At some point, other factors will become the rate-limiting step. This rate-limiting step in reinforced concrete is commonly the availability of sufficient oxygen.

1.4.2 Oxygen level

An essential factor for the corrosion of steel in concrete is the presence of oxygen at the steel to cement paste interface. The oxygen is required in addition to chloride or reduced alkalinity. If oxygen is not present, then there should not be any oxidation. For example, sea water has been used successfully as mixing water for reinforced concrete that is continually and completely submerged in sea water at the seabed. This is because of the maintenance of high alkalinity due to the sodium chloride (this boosts the concretes alkalinity due to the higher solubility of sodium ions in the cement paste) and low oxygen content in the sea water at the seabed and very slow diffusion rate of oxygen through the water-saturated concrete paste. There is initially a high corrosion rate when the critical chloride ratio was achieved. This depletes the available oxygen and then the corrosion rate dramatically reduces, despite an increasing chloride concentration. This slowing of the corrosion rate is assisted by a reduction in the oxygen solubility of water at very high chloride saturation levels, which further reduces the availability of oxygen. In most cases when the structure is submerged, the oxygen diffusion process is the rate-controlling step in the speed of the corrosion.

In chloride-free samples, when the pore saturation is reduced to 60% relative humidity a significant reduction in the corrosion rate is observed.

In chloride-containing samples, steel corrosion increases by about one order of magnitude (10 times) when reducing the pore saturation from 100% to 60%, obviously due to increased oxygen availability. Below 60% saturation, the corrosion rate tails off in a logarithmic manner until at 30% it becomes negligible.

The level of oxygen supply or resupply also has an effect on the corrosion products formed. A black product (magnetite) is formed under low oxygen availability, and a red–brown material (haematite) is favoured under high oxygen availability. The pore sizes of these oxides are different with the red product forming a more open structure with bigger pores. The formation of haematite imparts a higher bursting pressure on the concrete because of its greater volume and allows a quicker reaction to occur because of its greater porosity relative to magnetite. For these reasons, the presence of haematite rather than magnetite tends to indicate general corrosion rather than pitting and vice versa.

1.4.3 Cement type

Concrete composition has a significant bearing on the amount of corrosion damage that occurs at a chloride concentration. One example of this is that hardened concrete appears to have a lower chloride tolerance level than concrete that is contaminated during mixing. This is practically evident in precast units, which tend to corrode less than might be anticipated even when heavily dosed with a calcium chloride set accelerator (this is probably at least partially explainable due to their higher quality relative to cast-in-situ reinforced concrete of the same vintage and the absence of any potential differences caused by concentration gradients).

Although cement composition and type can affect corrosion, this effect is relatively small compared to the concrete quality, cover over the steel and concrete consolidation. Having said that, the use of a cement having a high C_3A content will tend to bind more chlorides and thus reduce the amount of chloride, which is free to disrupt the oxide film on the steel reinforcement. A cement of high alkali content would also appear to offer advantages because of the higher inherent alkalinity provided. In general it is observed that cements high in C_3A afford greater corrosion protection to reinforcing steel, but it is thought that other factors such as fineness and sulphate content may have at least as significant an effect. One study by Tuutti[9] found that Portland cement had a higher initiation level than slag cement but a lower diffusion resistance; thus, the study postulated that in certain exposure conditions a certain mix design with a Portland cement would be superior, whereas under other conditions the reverse was true and a slag cement would be superior. It was noted that a sulphate-resisting cement was always less effective than a Portland cement.

1.4.4 Aggregate type and other additives

In general, the higher the strength of the aggregate, the more likely it is to be resistant to the passage of ions. But this is not always so. For example, granite aggregate has been used for several major projects because of its high strength, but concrete made with this material has been found to provide relatively poor diffusion resistance results. This is probably because of micro-cracks in the aggregate or poor bonding between the cement paste and the aggregate.

It is likely that a substantial amount of the diffusion that occurs in concrete proceeds along the interface between the aggregate and the cement paste, and this region may well prove to be more critical than the bulk diffusion resistance of the aggregate. Certain aggregates have a smoother profile than others, and this will have an effect on the apparent diffusion path.

The addition of additives, such as microsilica, to concrete is beneficial as it increases the diffusion path by tending to block the pores in the concrete. With microsilica, the pH is reduced and thus corrosion may occur at a lower chloride concentration. A problem with this and other additives is the additional care required when it is being cast on the construction site and the assumption that with additives the concrete will change into a totally impermeable covering. This is not a safe assumption as the concrete will still retain a degree of porosity.

1.4.5 Temperature

The influence of temperature has a strong effect on the corrosion process of steel in concrete. It affects the corrosion potential, the corrosion rate, concrete resistivity and the transport process in the concrete.

The transport process and the electrolytic concrete resistivity strongly depend on the properties of pore water solution. The most important variable is the viscosity of water. Between 1°C and 50°C, the viscosity of water is reduced by a factor of more than three. As the viscosity is inversely related to the mobility of the particles, a change in temperature will be reflected closely by the transport rate. At a relative humidity of 80%, this transport mechanism is the rate-controlling step of the corrosion reaction and thus as expected this corrosion rate increases dramatically. However, the increase in rate is by a factor of more than four times, which points to additional factors favouring corrosion at higher temperatures. The ohmic concrete resistance and charge transfer resistance both show a temperature dependence that is similar.

There is a change in corrosion potential as the temperature is varied. This is normal for all aqueous electrochemical reactions. This change has been measured as a fall of 6.5 mV per 1°C[10] in salty concrete, whereas steel in a passive state has a fall of 2.5 mV per 1°C. This change in corrosion potential should have the effect of reducing corrosion rate as the temperature increases. However, practical experiments of corrosion rate make it apparent that this process has only a very limited effect on the kinetics of the reaction.

REFERENCES

1. A.A. Knudsen, *Trans. Amer. Inst. Elect. Engrs*, 26, 231, 1907.
2. P. Bamforth, *Concrete*, 18, Nov–Dec 1994.
3. S. Mattin and G. Burnstein, Detailed resolution of microscopic depassivation events in stainless steel in chloride solution leads to pitting, *Philisophical Magazine*, 76(5), 1–10, November 1997.
4. D. Hausman, Steel corrosion in concrete, *Materials Protection*, 6(11), 19–22, 1967.
5. V. Gouda, Corrosion and corrosion inhibition of reinforcing steel, I: Immersed in alkaline solutions, *British Corrosion Journal*, 5(9), 198–203, 1970.
6. V. Zivica, Corrosion of reinforcement induced by environment containing chloride and carbon dioxide, *Bulletin of Material Science*, 26, October 2003.
7. R. Podler, Reinforcement corrosion and concrete resistivity- state of the art, laboratory and field studies, Proceedings of International Conference on Corrosion Protection of Steel in Concrete, Volume 1, University of Sheffield, July 1994.

8. A. Bentur, S. Diamond, and B.N. Steven, *Steel Corrosion in Concrete*, Routledge, 1997.
9. K. Tuutti, *Corrosion of Steel in Concrete*, pp. 129–134, CBI Research, Stockholm, Sweden, 4.82, 1982.
10. S.E. Benjamin and J.M. Sykes, Effect of temperature and chloride content on the corrosion potential of iron in chloride contaminated concrete, *The Arabian Journal for Science and Engineering*, 20, 279–288, 1995.

Chapter 2

Corrosion in masonry structures

David Farrell

CONTENTS

2.1 INTRODUCTION

The mechanism of corrosion of metals when embedded in masonry structures (mainly iron and steel) is different from that of steel in concrete. The latter involves reinforcing steel and this is usually protected from corrosion for many years in good-quality, high-alkalinity concrete. The steel forms a very stable film of iron oxide on its surface, thereby mitigating any further corrosion. The stable film is called a passive layer and the protection is termed 'passivation'. For iron or steel embedded in masonry, a high-alkalinity environment is not always present. This depends on whether the metal is covered by a thin or thick mortar layer or, in many cases, fixed directly to the stone or brick. In the latter case, a passive layer will not be formed and corrosion may start soon after it is constructed. If the mortar layer is thin, or is of poor quality, then carbonation of the mortar may protect the metal for some years before corrosion initiates. In this chapter, the mechanism of corrosion is discussed. This is also covered in the Corrosion Prevention Association technical note number 20, *cathodic protection for masonry buildings incorporating structural steel frames*.

In the eighteenth and nineteenth centuries, dowels and cramps used in traditional masonry structures were usually made from wrought iron, which is susceptible to corrosion when exposed to air and moisture. The situation may be exacerbated if sedimentary stones, such as Portland and

Bath stones, are used because they frequently contain chloride and/or sulphate salts, which result in the depassivation of the iron surface and an acceleration of corrosion. Corrosion rates are significantly higher where iron is in direct contact with damp stone, rather than just moist air.

Some of the masonry-clad buildings that incorporate steel frames are also susceptible to corrosion. Steel corrodes at a higher rate than wrought iron in many situations, and this type of corrosion not only results in significant deterioration and loss of the original facade but also involves both health and safety issues, because of the risk of falling masonry, and involves costly and disruptive repairs. Conventional treatments can be highly invasive, involving large-scale opening up to expose and treat affected areas.

2.2 TRADITIONAL USE OF METAL FITTINGS

In major construction work dating from the eighteenth and nineteenth centuries, particularly porticos, arches and columns involving the use of full stones, large cramps, typically up to 1 m in length and 50 mm in thickness, were often surrounded by lead. Molten lead was poured into the gaps around the cramps to secure the iron fittings in place at the time of construction. Lead corrodes at a very low rate in this environment and if it completely surrounds the cramp, it should protect iron from corrosion for centuries. However, this is rarely the case: corrosion occurs on the non-leaded surfaces and progresses along the lead–iron interface (Figure 2.1). Eventually, expansion forces cause the lead to delaminate from the iron surface.

In ashlar masonry, where the stone layer is thin (typically around 50–100 mm) the smaller cramps have only limited cover (typically 20–30 mm). These cramps are not normally protected by lead, and it is common to find

Figure 2.1 Cramp with incomplete leaded surround at Dodington House.

vertical joints not filled with mortar to their full depth. When the shallow bead of mortar at the surface cracks or deteriorates, water can penetrate freely. The narrowness of the joints makes effective re-pointing very difficult, so water penetration continues, causing the embedded cramps to corrode, which results in the spalling of ashlar (Figure 2.2).

Whether unprotected or partly protected by lead, the expanding rust eventually exerts such pressure on the stone that the stone cracks or spalls. The volumes ratio between iron and rust can be as high as 1:7. Examples of spalled and broken full stones are shown in Figures 2.3 and 2.4.

Figure 2.2 Typical damage to the stone facade of the Whitchurch Almshouses due to expanding iron cramp.

Figure 2.3 Typical damage to the original stonework of the architrave at Dodington House caused by water ingress and corrosion of embedded cramps.

Figure 2.4 Damage to stonework at Great Witley church, Worcester, due to corroding cramps.

The conventional remedy for repair involves major surgery: removing the cramps, replacing them with non-corroding phosphor bronze or stainless steel and then repairing the damaged stonework. Cathodic protection offers an alternative approach to the treatment of rusting iron fittings and steelwork in masonry structures.

2.3 TRADITIONAL USE OF METAL REINFORCEMENTS AND SUPPORTS

Metals have been used for reinforcing and supporting masonry structures for many years. The example shown in Figure 2.5 comes from the fan vault at an English abbey. It boasts the earliest fan vault to be built in England (c. 1425). Located in the central tower, the fan vault had been subject to subsidence over the centuries and a wrought iron reinforcement system, covering the bottom and top faces of the ribs, had been installed in about 1840. Moisture ingress from the north and south walls had caused the iron reinforcement to corrode, ultimately leading to spalling of some of the stones and resulting in the tower being cordoned off in 2000.

Consideration was given to replacing the wrought iron with stainless steel reinforcement, but this was considered to be too disruptive and expensive. Besides, the Victorian era wrought iron itself was now of historic importance, and its replacement would have compromised the conservation principle of minimum intervention. Instead, a cathodic protection system was specified to provide protection to the iron reinforcement.

Figure 2.5 Damage to the fourteenth-century fan vault at Sherborne Abbey due to corrosion of the wrought iron reinforcements.

Figure 2.6 Damage to the eleventh-century Saxon stones in the vault at Gloucester Cathedral due to corrosion of the internal steel cores, which had been installed in the twentieth century.

An example of corrosion of steel support columns is shown in Figure 2.6. The crypt on this cathedral dates from the eleventh century and the stones are of historic value. Water logging has always been a problem in the crypt. In the early 1940s, the four eastern crypt piers were repaired by inserting steel columns into them. Because of the high levels of moisture within the crypt and the fact that the steel was not provided with a high-alkalinity mortar layer (the stone had been offered directly up to the steel), it started to corrode. This resulted in the expansion of the steel columns and cracking of some of the historic stones.

Consideration was given to dismantling the columns and drilling out the steel pillars, but this would have been very costly and intrusive and would have resulted in much damage to the historic stones. Again, cathodic protection was considered to provide the answer and a cathodic protection system was installed to suppress further corrosion of the steel.

2.4 MASONRY-CLAD STEEL-FRAMED BUILDINGS

Starting in the late eighteenth century and continuing into the early nineteenth century, iron- and steel-framed buildings were constructed in major cities throughout Europe and America. Thick load-bearing masonry-walled buildings, which had been common up to this time, were limited in their size of construction. The advent of steel-framed construction resulted in taller and lighter buildings than had been possible before. The period of masonry-clad building construction finished in the late 1930s when modern curtain wall designs allowed even taller and cheaper buildings to become possible. Some of the steel-framed buildings have been demolished over the past few decades, but the remaining examples represent a significant proportion of our historic buildings from this period.

In the early years of steel-framed construction, cast iron columns and wrought iron beams were used to support the masonry cladding. The cladding normally consisted of stone or brick, although glazed brick, Faience, and terracotta were also used. The wrought iron beams were replaced for 'new construction' in the 1890s as steel became more widely available. Finally, the cast iron columns were also replaced with steel at the end of the nineteenth century.

For most of the steel-framed buildings, the large facade stones were cut to fit closely to the steel framing. The gaps were normally filled with poor-quality mortar, sometimes containing brick and rubble aggregate (Figure 2.7). This allowed moisture to accumulate within the cementitious rubble, which was in direct contact with the steel. The architects at the time possibly thought that the alkalinity of the infill would be sufficient to passivate the steel and prevent corrosion, in a similar manner to that experienced for concrete. However, the poor quality and porous nature of the fill gave only limited protection to the steel framing and carbonation of the mortar–rubble infill resulted in depassivation of the protective oxide surface film on the steel. This, in combination with the accumulation of moisture within the infill, resulted in corrosion of the steel framing.

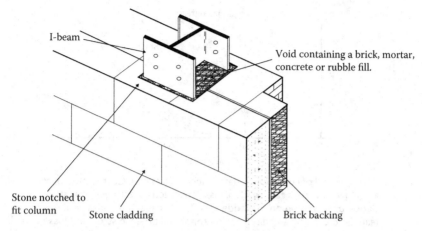

Figure 2.7 Schematic diagram of masonry-clad steel-framed construction detail.

2.4.1 Corrosion mechanism

Corrosion of steel frames in masonry-clad steel-framed buildings is the single and most costly problem facing owners of these buildings today. Even low levels of corrosion are sufficient to crack or spall stone facades because of the volumetric expansion of steel as it is converted to a corrosion product. The expansion forces result in internal stresses building up within the walls, which results in initially cracking followed by 'jacking out' of the stones. Once this damage starts to occur on the masonry, the damaged areas allow further ingress of moisture, which often results in an acceleration of corrosion and the worsening of deterioration.

Corrosion is an electrochemical process and both oxygen and moisture are required for it to proceed. Iron and steel in either very dry conditions (no moisture) or totally submerged (no oxygen) is not subject to corrosion. The actual proportions of oxygen and water present on iron or steel surfaces determine the rate of attack. Oxygen is normally present in sufficient quantities in trapped air, and increased moisture normally results in increased corrosion.

The following types of corrosion are frequently found on masonry-clad steel framing:

Uniform (or general) corrosion: this is frequently found as a general type of corrosion or rust covering metal surfaces. It is the most common form of corrosion and is normally attributable to carbonation of the mortar infill, which results in depassivation of the protective oxide layer.

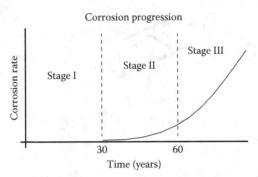

Figure 2.8 Staged model of corrosion initiation and progression. Corrosion progression is as follows: stage I is carbonation, loss of protection to steel frame; stage II is initiation of corrosion on embedded steel; and stage III is widespread corrosion, leading to cracked and displaced masonry and possible structural loss of steel.

Pitting, or localised corrosion: this occurs in localised areas only, but results in high rates of attack. It is generally uncommon for masonry-clad steel construction. However, it sometimes occurs where water ingress is localised to a small area or in coastal environments where chloride ions, from marine rainfall or salt spray, allow a build-up of chlorides to occur at steel surfaces. The chloride ions depassivate the steel, which results in high levels of attack at selected locations.

The carbonation of any mortar layer covering the steel frame is similar to that for concrete. However, the rate of carbonation, and thus the time to initiation of corrosion, is dependent on the quality of concrete and its thickness. For the poor-quality mortar infill that was normally used for this type of construction, the time to initiation might be only a few decades. A general description of carbonation and time to initiation and progression of corrosion is given in Figure 2.8. Carbonation of a thin layer of poor-quality mortar takes around 30 years, after which depassivation of the protective oxide layer occurs and corrosion initiates. This continues for a while, but after around 50 years the corrosion attack starts to accelerate as the masonry becomes cracked and this allows increased moisture to enter the construction. If the steel framing is fixed directly to the masonry and is not protected by a mortar layer, then the stage I process (carbonation) can be omitted and stage II will begin soon after construction.

2.4.2 Examples of corrosion of steel frame construction

A schematic diagram of the typical corrosion of a steel I-beam and the resulting fracture of a stone facade is given in Figure 2.9. Moisture generally enters the construction and a build-up occurs towards the bottom

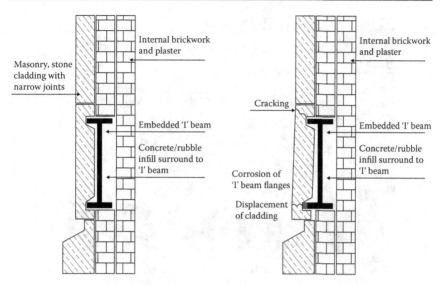

Internal brickwork and plaster

Masonry, stone cladding with narrow joints

Cracking

Embedded 'I' beam

Concrete/rubble infill surround to 'I' beam

Corrosion of 'I' beam flanges

Displacement of cladding

Internal brickwork and plaster

Embedded 'I' beam

Concrete/rubble infill surround to 'I' beam

Figure 2.9 Schematic showing the build up of moisture in steel framed construction and consequent stone damage.

Figure 2.10 Practical example of the damage to steel framed construction as illustrated in Figure 2.9.

flange surface of the beam. Corrosion ensues and the expansion forces a crack and then pushes out the facing stone. A practical example of the damage that this can cause to the facade is shown in Figure 2.10. Removing

Figure 2.11 Opening up the steel framed construction to reveal the corroding steel beam.

Figure 2.12 Typical corrosion of steel framing from twentieth century.

the outer stone reveals the corroding I-beam below (Figure 2.11). Another example is shown in Figure 2.12. In this case, the corrosion on the steel framework was general and widespread and was due to moisture ingress through the joints in the corners of the facade. Note: moisture may enter into the construction many metres away from the areas of damage.

Figure 2.13 Close up of expanding corrosion layer that results in jacking-up of the masonry.

The expansion forces resulting from corroding iron and steel are extremely high, and it is reported that the corrosion of a metal ring beam at St. Paul's Cathedral, London, has raised the entire central dome of the building, itself weighing many thousands of tons. An example which resulted in the lifting of the front of a roof on a large country house is shown in Figure 2.13. Moisture ingress some metres away resulted in the corrosion and expansion of a steel wall beam. The corroding web of the I-beam has pushed up the stones and roof structure by around 20 mm.

An example of cracking of a stone pillar, because of ongoing corrosion and expansion of the internal steel frame, is shown in Figure 2.14. Moisture entered the pillar because of faults in the roof construction and ran down the inside of the pillar and built up in the mortar layer separating the steel from the stone.

A further example of corrosion to a steel column with brick cladding is shown in Figure 2.15. The building was constructed in the 1940s, and the steel had been covered with a 'red lead' coating to protect against corrosion. Note: coatings contain 'holidays' or defects, which allow a small amount of moisture to penetrate down to the steel. Coatings can only slow down the time to initiation of corrosion; they cannot fully stop it. The coating had deteriorated (because of undermining corrosion) where the steel was in direct contact with the mortar and brick and ongoing corrosion and expansion had cracked the outer brick. The 'red

Figure 2.14 Cracking of the vertical pillars to a cricket pavilion at Downside Abbey and School, Somerset.

Figure 2.15 Damage to a vertical brick column due to corrosion and expansion of the steel frame.

lead' coating on the column, which had not been in direct contact with the damp surfaces and had been exposed only to air, remained intact.

An example of corrosion to a wrought iron framework is shown in Figure 2.16. This 1858 building was extensively refurbished in 2001. The

Corroding support
beams

Figure 2.16 Cracking of the statuary at the Royal West of England Academy, Bristol, due
to corrosion of the steel framing.

pediment was constructed of clinker concrete with a render finish and had
I-beams within the central construction to provide support. Clinker, a by-
product of coal combustion in power stations, contains significant quantities of chlorides, which, in combination with water ingress through the
render, resulted in corrosion of the iron beams and cracking of the concrete. The front of the pediment contained various pieces of statuary that
had been cast into the outer stones and this important facade would have
been seriously damaged if the steel had had to be replaced. It was therefore
decided to use a cathodic protection system to control the corrosion of the
embedded iron beams.

Chapter 3

Site appraisal to enable efficient cathodic protection design

John Broomfield

CONTENTS

The appraisal of a structure for its suitability for the application of cathodic protection will be part of the process of a general condition assessment and repair design. The first aim of that process will be to ensure the safety of the structure, in terms of both structural integrity and the risk of anything falling off the structure because of deterioration. Health issues such as the presence of asbestos and its disturbance during the repair work may also require assessment.

After the health and safety issues, the most important criterion is to determine the cause or causes of deterioration so that appropriate repair options are considered. For cathodic protection to be appropriate, the cause of deterioration must be reinforcement corrosion. Any other causes of deterioration must also be adequately addressed in the repair strategy. There are several guidance documents on the assessment of reinforced concrete structures.[1,2] Some documents are specifically related to reinforcement corrosion.[3,4] These give references to European and U.S. standard test methods for measuring the chloride content of concrete, carbonation depths, delaminations and other test methods, which will not be considered here.

Once corrosion is established as the primary cause of damage and the extent has been quantified, the following criteria can be applied to determine eligibility for cathodic protection:

1. The extent of deterioration and remaining life of the structure
2. The type of reinforcement
3. Presence of pre-stressing
4. The type of concrete
5. Availability of electrical power
6. Availability of telecommunications
7. Suitability of available anodes
8. Electrical continuity of the steel
9. Other special conditions

Although there are standards for concrete repair that list the available options such as the new European BS EN 1504-9,[5] they give few technical criteria for selecting the optimum repair and corrosion control method. One criterion that is mentioned is life-cycle cost analysis. It is a standard requirement for most U.K. and many other European government projects that the repair options are listed with their strengths and limitations and their relative costs in terms of first cost and life-cycle cost. However, to compare costs, the engineer needs access to the relative unit costs of the activities and materials involved in each repair option.

Standardised unit pricing for different repair costs were developed for a U.K. Department of Industry project, and methods for predicting time to corrosion, rate of deterioration and life-cycle costing are given on the BRE website.[6] An update of costs for impressed current cathodic protection of U.K. highway bridges in given in CPA Technical Report 12.[7] However, these costs are a considerable increase on those given in the BRE website.[6] This is thought to be due to contracting and access costs for cathodic protection of highway structures, which are different from most other projects and difficult to separate out.

One of the advantages of life-cycle costing is that it requires the structure owner to give some thought to how long he or she wants it to last; what level of deterioration is acceptable; and whether, when and what types of further cycles of major and minor interventions are acceptable. A cycle of patching up a structure every 5 years may be acceptable for something with a limited life or with limited funding, but eventually it will become irreparable. Some structures even with a nominal life of 50 or 100 years may, in fact, be required to last indefinitely, so any deterioration should be minimised and leave the structure sound after all foreseeable rounds of repair. Impressed current cathodic protection with long-life anodes may be the most technically sound method with the lowest life-cycle cost in these situations. Structures in the early stages of corrosion or where problems are localised may be treatable with local repairs and protection with coatings.[8]

As well as the condition survey, it is essential to carry out a desk study of the available design drawings and maintenance history of the structure. It is also essential to understand the client's requirements for the structure and any parameters that may affect the installation and performance of repair systems.

However, the relative costs of repairs are irrelevant if they are technically inappropriate, as discussed in the remainder of this chapter.

3.1 EXTENT OF DETERIORATION OF A STRUCTURE AND REMAINING LIFE

Impressed current cathodic protection is unlikely to prove cost-effective on a structure with less than a decade of remaining life once comparative life-cycle cost analysis is carried out. Patch repairs combined with anti-carbonation coatings can last 10 years on a carbonated structure. On structures subject to chloride-induced corrosion and chloride ingress, repairs are likely to fail and new areas of damage emerge in 3–5 years depending on the severity of the environment. However, galvanic anodes may be suitable to protect repairs and stop new corrosion outbreaks.

The amount of repair required does not impact the technical feasibility of cathodic protection, as most structures will require repairing. However, it may impact the cost-effectiveness in comparison with partial or full replacement. Also, the existence, type and condition of previous repairs may have an impact on the project. Certain types of repairs such as epoxy mortars are incompatible with impressed current cathodic protection.[9]

The advantage of cathodic protection is that it can control corrosion of the whole of the area treated. Impressed current cathodic protection can control corrosion regardless of chloride content in concrete and ongoing chloride ingress. It minimises the amount of concrete repair required as repairs are providing a current path to the steel, not removing chloride contamination. There is therefore a lower requirement for breaking out and replacing concrete and, therefore, any requirement for structural propping during repairs can also be reduced or eliminated.

3.2 TYPE OF REINFORCEMENT

All common types of bare steel reinforcement are candidates for both impressed current and galvanic cathodic protection. Coated reinforcement, such as fusion bonded epoxy coated reinforcement, has been the subject of large numbers of cathodic protection installations in the Florida Keys.[10] It is important to ensure that all reinforcement is electrically continuous, particularly for impressed current systems, and this can be done in a cost-effective manner. This author is not aware of any cases of cathodic protection of galvanised reinforcing steel or stainless steel–reinforced structures, but in principle they can be protected in the same way as plain reinforcement.

3.3 PRESENCE OF PRE-STRESSING

The presence of pre-stressing means that a very careful assessment is required. The design drawings need evaluation as well as any survey data. The questions to be asked are as follows:

- What is corroding, conventional reinforcement, steel ducts around the pre-stressing or pre-stressing itself?
- If it is the pre-stressing itself, can a cathodic protection system pass current and protect the vulnerable steel?
- If the pre-stressing is not corroding, can the vulnerable steel be protected from the risk of hydrogen embrittlement?
- Is the pre-stressing steel susceptible to hydrogen embrittlement?
- Can galvanic cathodic protection be applied? If so, the risk of hydrogen embrittlement becomes negligible.
- If impressed current cathodic protection is necessary and there is a risk of hydrogen embrittlement, can it be adequately monitored and controlled?

There is a very comprehensive report on the impressed current cathodic protection of pre-stressing published by NACE.[11]

3.4 TYPE OR CONDITION OF CONCRETE

Most normal concretes that will support corrosion will transmit sufficient current from a suitably chosen anode system to vulnerable reinforcement. Exceptions might be porous concrete that is very dry most of the time but is wetted occasionally. This author has heard of a situation where a conductive coating anode was applied to the soffit of a heated multi-storey car park where cars brought in salt-laden snow and ice, which soaked into the top cover but the anode was unable to pass sufficient current from the soffit to protect the steel because the soffit concrete was so dry. In this case, the positioning of the anode led to the unforeseen problems.

The existence of reactive aggregates in concrete has given rise to much laboratory research but very few field investigations. Alkali–silica reaction (ASR) is caused by reactive aggregate particles reacting with the alkalinity in concrete. This process is now well understood and addressed during concrete mix design by testing aggregates and controlling the maximum alkali content of cements.[12] In principle, the alkali generated at the steel surface by the cathodic reactions could cause ASR. In practice, where older structures have been found to have either active ASR or the potential for ASR and have been treated with impressed current cathodic protection, no evidence of generation or exacerbation of ASR has been observed.

3.5 AVAILABILITY OF ELECTRICAL POWER

Impressed current cathodic protection systems and automated and remote monitoring systems require electrical power. Although this is usually from the electrical mains grid system, alternative supplies such as wind turbines, photovoltaics and batteries have been used. Unfortunately, they are always used in remote sites and therefore can be vulnerable to vandalism and theft. There are also cost and maintenance implications for these alternative power sources. Even impressed current systems do not need a guaranteed, 100% supply; so the occasional loss of power because of lack of wind and/ or sun is acceptable, although it can complicate the interpretation of performance data.

3.6 AVAILABILITY OF TELECOMMUNICATIONS

Many impressed current cathodic protection systems are remote controlled and operated as discussed in Section 3.10.4. These benefit from having an available dedicated landline. However, where a direct telephone landline is unavailable, cell phones or satellite communications have been used. Data transfer rates do not need to be particularly rapid.

3.7 SUITABILITY OF AVAILABLE ANODES

There is now a wide range of different types and sizes of anodes available, especially for impressed current cathodic protection. These can usually be installed unobtrusively on most elevations of most reinforced concrete structures. For galvanic anodes, the requirement to have sufficient material available to corrode sacrificially and give a reasonable anode life means that they may be more obtrusive and invasive than impressed current anodes. The different types of anodes currently widely available and their capabilities and limitations are discussed in Chapter 5 and elsewhere.[4]

3.8 ELECTRICAL CONTINUITY OF STEEL

It is essential that all steel to be protected is electrically continuous and connected to the negative terminal of the power supply. All other embedded steel likely to receive current from the anode (say, within about 0.5 m from the edge of the anode zone) will usually need to be made electrically continuous with the protected steel in an impressed current cathodic protection system. This requirement can be relaxed in a galvanic cathodic protection system as stray currents are less likely to occur and less likely to be damaging.

3.9 OTHER SPECIAL CONDITIONS

Each structure is unique and has its own requirements. It is therefore difficult to cover all issues in a generic review such as this chapter. Historic listed structures have their own requirements, some of which will be general, as discussed in Chapter 2 on the cathodic protection of steel-framed buildings, and some of which will be particular because of the particular structure's history, construction or appearance.

Care must be taken in the presence of waterproofing. Impressed current anodes generate oxygen and chlorine gas. These must be able to escape. If there is waterproofing or saturated concrete, it is possible for gas pressure to build up around the anode and cause localised damage and anode failure.

It is also important to consider any other deterioration processes, possible side effects and interactions.

3.10 DESIGN ISSUES

During the assessment process, it is usually necessary to consider certain design issues such as the following:

1. Total current demand
2. Anode type
3. Zoning
4. Power supply and control system type and location

3.10.1 Total current demand

The total or maximum design current demand is calculated by multiplying the surface area of the steel to be protected by the maximum design current density. This requires information on reinforcement spacing and bar diameters. This may also require deciding how many layers of steel will receive current. A simple rule of thumb is that each successive layer receives 30% of the layer above it. It also requires deciding on the design current density. The design current density limits for the cathodic protection of new, of non-corroding and of existing corroding structures are given in the non-mandatory Annexe 1 of BS EN ISO 12696.[13] It should be noted that the repair process will remove large areas of active corrosion so that the current demand may be lower than the maximum 20 mA·m^{-2} recommended in the standard. Most systems run at far lower currents than their design currents.

3.10.2 Anode type

Once the maximum current is calculated, or at least estimated for initial assessment purposes, suitable anode types are assessed. Chapter 5 describes the available anode types in detail, but some might be excluded for reasons of inadequate life (e.g., conductive organic coatings), increase in load (e.g., overlay systems) or other constraints such as appearance or ease of installation. Most anodes come in a range of current capacities, and the selected anode type or types must be capable of delivering the required current density to steel.

3.10.3 Zoning

Most cathodic protection anode systems are divided into zones. For large monolithic structures such as a bridge or a car park deck, the decision may be based on simple geometry determined by the current output from the modular power supply in the case of impressed current systems. On more complex structures, it will be based on the design of the elements to be protected, along with exposure conditions and variations in steel surface area per unit concrete surface area and the anode types under consideration.

3.10.4 Power supply and control system type and location

It is important to decide at an early stage how the system is to be monitored and controlled, particularly for impressed current systems. Larger systems with multiple zones and monitoring sensors will benefit from automated operation and logging of data. In most such cases, they will also benefit from being able to communicate from a remote location to download data, troubleshoot and adjust the system as discussed in Chapter 10.

On larger, more complex impressed current systems, power supplies and local data loggers can be distributed throughout a structure to ensure that they are close to the zones they control with a suitable network to a central processor unit at a convenient location for power, data storage and telecommunications.

Small systems with few zones in easily accessed locations can have a simple manual monitoring and control system where an operative goes to the site, takes current and voltage readings, measures reference electrode potentials and depolarisations and adjusts the system (for impressed current systems) at a single control and monitoring location.

The location of enclosures for the electronic and electrical systems must also be considered in terms of both environment and cable runs. In some cases, they may be in a climate-controlled secure room, whereas in others they may be in a corrosive outdoor exposure condition with risk of damage and vandalism. The types of enclosures used to house the electronics will need to be suitably specified for the environment. It can be embarrassing to try to open an enclosure box housing a state-of-the-art cathodic protection system and find that the locks and hinges have rusted solid.

3.11 SUMMARY

There is a range of options for rehabilitating damaged reinforced concrete structures. A proper assessment of the structure will determine which are technically suitable. This will depend on issues including health and safety, the actual cause of deterioration and the ongoing demands on the structure.

Life-cycle costing can be carried out to determine the most cost-effective rehabilitation solution. If cathodic protection is chosen, then the first choice is between impressed current or galvanic anode systems. Galvanic systems have no control systems and a lower level of monitoring and cannot be guaranteed to control corrosion. Anode life is typically limited to 20 years or so. Impressed current systems can provide complete corrosion control across the whole area treated but require a higher level of ongoing maintenance. Anode life can be 100 years or more.

Specific issues of design, construction, and condition of the structure should be considered, especially for impressed current cathodic protection. However, these are more likely to impact the cost of the installation work rather than the viability of applying cathodic protection.

REFERENCES

1. Concrete Society (2000) Technical Report 54, *Diagnosis of deterioration in concrete structures*, The Concrete, Camberley, Surrey, UK.
2. Concrete Bridge Development Group (2002) Technical Guide 2, *Guide to the testing and monitoring the durability of concrete structures*, Publ. Concrete Society, Camberley, Surrey, UK.
3. NACE (2008) *SP 0308 Inspection methods for corrosion evaluation*, NACE International, Houston, TX.
4. Broomfield, JP (2007) *Corrosion of steel in concrete—understanding, investigation and repair*, 2nd Edition, Taylor and Francis, London, UK.
5. BS EN 1504-9 (2008) *Products, and systems for the protection and repair of concrete structures—Definitions, requirements, quality control and evaluation of conformity—Part 9: General principles for the use of products and systems*, British Standards Institute, London, UK.
6. Rebar Corrosion Cost Website. *Building research establishment.* http://www.projects.bre.co.uk/rebarcorrosioncost/ (last accessed 2 June 2011).
7. CPA Technical Note 12 (2008) *Budget cost and anode performance information for impressed current cathodic protection of reinforced concrete highway bridges*, Corrosion Prevention Association, Farnham, Surrey, UK.
8. Broomfield, JP (2006) *A web based tool for selecting repair options and life cycle costing of corrosion damaged reinforced concrete structures* Proc. Concrete Solutions Conf. Eds. MG. Grantham, R Jaubertie and C Lanos, Publ. BRE Press, Watford, Hertfordshire, UK.
9. CPA Technical Note 19 (2011) *Acceptable electrical resistivities of concrete repairs for cathodic protection systems* Corrosion Prevention Association, Farnham, Surrey, UK.
10. Kessler, RJ, Powers, RG, Lasa, IR (2002) *An update on the long term use of cathodic protection of steel reinforced concrete marine structures* NACE Corrosion 2002 Paper No. 02254, NACE International, Houston, TX.
11. NACE Report 01102 (2002) *State-of-the-art Report: Criteria for the cathodic protection of prestressed concrete structures.*
12. BRE Digest 330 (Parts 1 to 4) (1999) *Alkali-silica reaction in concrete*, Building Research Establishment, Watford, Hertfordshire, UK.
13. BS EN ISO 12696 (2012) *Cathodic protection of steel in concrete*, British Standards Institute, London, UK.

Chapter 4

Cathodic protection mechanism and a review of criteria

Kevin Davies and John Broomfield

CONTENTS

4.1 INTRODUCTION

The mechanisms of corrosion of steel in concrete are discussed in Section 4.3. In this chapter, they will be reviewed while considering how cathodic protection (CP) can be used to control corrosion. We review the electrochemical potentials and the corrosion currents of corroding and 'cathodically' protected steel in concrete and then review the criteria for achieving CP and how these are reflected in the major standards for CP of steel in concrete currently available.

4.2 PASSIVATION

Reinforced concrete is a very commonly used, high volume building material and provided the steel reinforcement, required to provide the tensile capabilities; is well encased in non-contaminated, well-compacted concrete with adequate cover; it is very versatile and very durable even in the harshest of environments. The concrete matrix provides a good physical and chemical protective barrier to the steel.

The presence of calcium, sodium and potassium hydroxides produced during the cement hydration (concrete curing) reactions produces a highly alkaline (pH 12–13) environment surrounding the steel reinforcement. The pores in the cement matrix contain water with saturated hydroxide solutions and there is excess solid calcium hydroxide in particular that will go into solution if the concentration drops. This is very important both for corrosion protection and for CP. The hydroxides in solution react chemically with the surface of the steel to form a thin (believed to be only some 2–3 nm thick) dense, protective oxide passive layer composed mainly of iron hydroxides. The presence of this 'passive layer' reduces corrosion of the steel to negligible levels (<1 µm metal loss per year). In stable, alkaline environments, the steel is considered to be 'passive' and even small breaks in the protective oxide film are repaired efficiently by the buffering action of the alkaline reservoir within the pore water. Corrosion is controlled to negligible levels.

There is still a limited understanding of the exact nature of the passive oxide layer developed on steel surfaces in alkaline concrete. Some researchers have indicated that the passive layer is very complex and the oxide layer is composed of various forms of $Fe(OH)_2$, $Fe(OH)_3$, Fe_2OH_3 and Fe_3O_5. It is most often described as $\gamma FeOOH$ (Bentur, Diamond, and Berke 1997).

4.3 CORROSION MECHANISM

Steel is a man-derived material created by taking natural iron ore and injecting vast amounts of energy to produce usable malleable, formable and weldable metal. The smelting process involving extensive heating places the formed metal into an artificially high-energy state. Corrosion is simply the natural, spontaneous, exothermic process of the iron trying to revert to its natural lower energy state (oxide) form. No further energy input is required for corrosion, merely the presence of oxygen to react with and, in normal atmospheric conditions, moisture to mobilise the reactants and corrosion products.

The protective passive oxide layer can be broken down in two ways without any apparent damage to the concrete. One is by atmospheric carbon dioxide (CO_2) reacting with atmospheric moisture to produce carbonic acid, which neutralises the alkalinity in the pores. Once the carbonation front reaches the steel, the passive layer is no longer maintained and breaks down. The other mechanism is chloride ion (Cl^-) attack from marine environments, cast in additives and in the United Kingdom from road salt. The chloride ions react with the hydroxyl ions forming and maintaining the passive layer, and once the critical concentration is exceeded, the protective passive layer breaks down. Once the passive layer is breached, steel will corrode naturally in moist environment due to differences in the electrical potential on the steel surface forming anodic (corroding) and cathodic (protective) sites.

Iron oxidises (corrodes), releasing electrons (e^-) into the metal at the anodic sites.

$$Fe \rightarrow Fe^{2+} + 2e^- \qquad (4.1)$$

These electrons cannot build up in the steel, so a reduction reaction occurs at cathodic sites. A typical cathodic reaction is

$$H_2O + \tfrac{1}{2}\,O_2 + 2e^- \rightarrow 2OH^- \qquad (4.2)$$

However, another cathodic reaction can occur if the potential gets very negative:

$$H_2O + e^- \rightarrow H + OH^- \qquad (4.3)$$

As long as the concentration of ions in solution (e.g. Fe^{2+}, OH^-) has balanced electrical charge, corrosion can proceed.

We can therefore see that we have an electrical circuit and we can apply Ohm's law to the corrosion current and electrochemical potentials or voltages that characterise the corrosion cell.

According to Ohm's law ($I_c = \Delta E/R$), this can be simplified to

$$I_c = \frac{E_c - E_a}{R_c + R_a}$$

where,

I_c is the overall corrosion current (mA)
E_c is the cathode potential (mV)
E_a is the anode potential (mV)
R_c is the resistance at the cathode (ohms)
R_a is the resistance at the anode (ohms)

This is a simplified equation but does demonstrate some important concepts.

When $E_c = E_a$, that is, when there are no sites with different potential on the surface of the steel, there is no corrosion as $I_c = 0$.

Figure 4.1 shows a simplified potential versus corrosion current diagram. If there is no electrical resistance in the circuit, the theoretical measured potential will be the average of E_c and E_a. In this case, the corrosion current is the maximum that the coupled anodic and cathodic reactions can achieve. However, in practice, the electrolyte (concrete) will provide a circuit resistance and the corrosion current will be reduced and we will be able to measure potentials that approach of E_c and E_a as shown by the double-headed arrow. In good quality concrete with a continuous passive layer on the steel, the resistance is very high and so the current is very low.

One problem, from a practical point of view, is that it is not possible to determine the locations of anodic and cathodic sites accurately. It is not possible to measure the potentials just at the *anodic* and the *cathodic* sites as even in concrete there is likely to be a multitude of these sites and they may be very small (pits) or very close to each other and often beneath some 50 mm of cover concrete. In many cases, they are also not static and can move around as environmental conditions change.

Using an external reference electrode (half-cell) on the concrete surface, the combined polarised potential E_m can be measured directly. The measured potential (E_m) will be somewhere between E_c and E_a. Potential

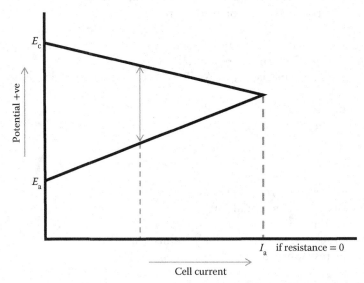

Figure 4.1 Schematic of anode and cathode potentials versus corrosion current where the anode and cathode resistances are equal.

measurements from the concrete surface merely produce gross (weighted) average mixed potentials (E_m) of the steel.

$$\left(E_c > E_m > E_a\right)$$

From the simplified diagram in Figure 4.1, the actual potentials at the cathode (E_c) and anode (E_a) cannot be measured independently as they are not electrically isolated from each other by being electrically connected by the steel rebar. Hence, they tend to move towards each other. A reference electrode placed on the concrete surface will only enable measurement of a mixed steel potential (E_m), which is somewhere between (E_c) and (E_a).

ASTM C876 provides guidelines for assessing steel corrosion based on measured corrosion potentials. The guidelines suggest that for atmospherically exposed concrete steel, potentials less than -0.150 V with respect to silver/silver chloride/potassium chloride (Ag/AgCl/KCl) indicate a very low (<5%) risk of corrosion. NACE RP 0290 also suggests that steel with measured potentials in this range do not require the application of any CP current as the steel would be considered to be in a passive state.

Just as we cannot measure the actual potentials of the anodes and cathodes, it is also impossible to directly measure the current between anode and cathode. However, there are ways of measuring it indirectly.

Applying an external negative potential to the steel will force the measured potential E_m become more negative until at some point, when the applied current equals I_p the measured potential E_m will reach E_a. At this point, E_m + (the applied negative potential) will equal E_a. This implies that E_c will have been polarised negatively to E_a and at this point there will be no natural driving potential and the natural corrosion process will stop. There will be no net corrosion current flowing between the original anodic and cathodic areas on the steel surface, so the corrosion will have been mitigated. This is the principle of applied CP. If the negative charge can be applied early in the structure's life, while the passive layer is still in place and before corrosion currents have initiated, it is termed 'cathodic prevention', which is basically just a way to supplement the passive layer formation.

For most existing reinforced concrete structures, the concrete will have already cured to a great extent and with environmental exposure the steel reinforcement would normally be at its natural corrosion potential. At this point, the relationship between applied current and potential change is almost linear. Corrosion monitoring devices use this principle to estimate corrosion rates of steel in concrete. One example is where a small perturbation current is applied to the steel using an auxiliary electrode causing an anodic potential shift of ~20 mV. The current required to do this is measured and using accepted constants and the Stern–Geary equation corrosion rates can be estimated per unit area of steel (Broomfield, 2007).

Corrosion rates for steel in concrete are usually quite low when compared to rates seen for exposed steel in atmospheric or marine conditions. Corrosion rates range from 0.1 $\mu A/cm^2$ (1 mA/m^2) indicating passive conditions through to 1 $\mu A/cm^2$ (10 mA/m^2) indicating active corrosion. Some exceptionally high corrosion rates of up to 10 $\mu A/cm^2$ (100 mA/m^2) are possible.

We can convert corrosion currents to corrosion rates using Faraday's equation of electrochemical equivalence:

$$\text{Metal loss rate} = I_{corr}/\text{Area} \times \text{Time} \times \text{Atomic Weight}/(F \times \text{Valency})/\text{Density}$$

where,

I_{corr} = Corrosion rate ($\mu A/cm^2$)
Area = cm^2
Time = $365 \times 24 \times 60 \times 60$ (seconds)
Atomic weight = 55.85
$F = 96.485$ (A.s)
Ionic charge = 2 (for iron)
Density = 7.87 (g/m^3)

For steel with a corrosion rate of 1 $\mu A/cm^2$, the theoretical calculated metal section loss would be 11.6 $\mu m/year$.

Table 4.1 lists commonly quoted general corrosion currents and their basic interpretation (Broomfield, 2007).

If an uncoated steel component was left exposed in a marine environment, anodic and cathodic sites would form on the steel surface and in time these will cover the exposed surfaces. The anodic and cathodic sites will also move around as conditions change and although corrosion only ever occurs at the anodic sites an appearance of all over 'general' corrosion would soon be seen. With steel embedded in concrete, the anodic and cathodic sites are more defined as anodic sites tend to stay in their location, being more dependent on variations and defects in the concrete or on the local environment than variations in the steel surfaces.

Table 4.1 Corrosion currents—Steel in concrete

Corrosion current ($\mu A/cm^2$)	Corrosion current (mA/m^2)	Corrosion rate
<0.1	1.0	Passive—very low corrosion
0.1–0.5	1.0–5.0	Low to moderate corrosion
0.5–1.0	5.0–10.0	Moderate to high corrosion
>1.0	>10.0	High corrosion
1.0–10.0	10–100	Very high corrosion

4.4 CATHODIC PROTECTION MECHANISM

If we purposely introduce additional anodes in, or onto, an atmospherically exposed reinforced concrete structure, we can pass a direct electrical current to the reinforcing steel to control corrosion. With the new anode charged positively and the reinforcing steel charged negatively, we can see that the anodic reaction 4.1 will be suppressed as the negative ferrous ions (Fe^{2+}) cannot escape the negatively charged steel and the cathodic reaction 4.2 occurs. The production of OH^- ions helps to re-alkalise the area around the steel and if it is carbonated helps to restore and maintain the passive layer. In addition, by charging the steel negatively, the negatively charged chloride ions (Cl^-) will be repelled away from the steel and attracted to the positively charged anode. This helps prevent the passive layer from being attacked.

Returning to Figure 4.1, applying an externally produced negative potential to the steel will force the cathode potential E_c to become more negative, until at some point it will reach E_a. At this point, there will be no natural driving potential between anode and cathode on the steel and the natural corrosion process will stop. There will be no net corrosion current flowing between the original anodic and cathodic areas on the steel surface, so the corrosion will have been mitigated.

The relationship between applied CP current and the shifts in potential produced is quite complex and depends on a number of parameters including steel depth, quality of concrete, mix design, CP anode type and distribution, moisture and oxygen content among others. Each application of CP to steel in concrete will, therefore, have its own design CP parameters and performance characteristics.

In addition to the beneficial reactions that occur on the steel surface, there is one deleterious one. This hydrogen (H) evolution reaction (4.3) can lead to hydrogen embrittlement in exceptional circumstances. Monatomic hydrogen is evolved at the steel surface. This can combine to form hydrogen gas that very easily escapes through the concrete. However, it can also diffuse into the steel, and in certain susceptible steels, it can get trapped at grain boundaries or defects in the crystalline matrix of the steel, weakening it and causing failure under load. This problem is negligible for normal reinforcing steel, but is of considerable concern for pre-stressed structures where the certain high-tensile steels can be susceptible to hydrogen embrittlement and can affect its ultimate tensile strength leading possibly to catastrophic failure.

The problems of hydrogen embrittlement and of gas evolution are usually controlled by limiting the negative potential of the steel to above the hydrogen evolution potential. However, within pits or crevices, the potential can exceed (be more negative) the hydrogen evolution potential without being sensed by measuring electrodes. The CP of pre-stressed structures should only be undertaken with great care and input from experienced corrosion experts. A state-of-the-art report on the CP of pre-stressed has been published (NACE, 2002), which explains these issues in great detail. To avoid excessive hydrogen gas evolution, a potential limit is set, even for conventional reinforcement with no susceptibility to hydrogen evolution.

As well as considering the chemical reactions at the cathode, we should also consider those at the anode. The chloride ion itself is negative and will be repelled by the negatively charged cathode (reinforcing steel). It will move towards the (new external) anode. With certain types of anodes, it may then release its electric charge to the positive anode and form chlorine gas:

$$2Cl^- \rightarrow Cl_2 \, (gas) + e^- \tag{4.4}$$

The other major reaction at all major anodes is the formation of oxygen:

$$2OH^- \rightarrow H_2O + \frac{1}{2}O_2 + e^- \tag{4.5}$$

and

$$H_2O \rightarrow \frac{1}{2}O_2 + 2H^+ + 2e^- \tag{4.6}$$

Reaction 4.5 is the reverse of reaction 4.2, that is, alkalinity is formed at the steel cathode (enhancing the passivity of the steel) and consumed at the anode. These and related reactions can carbonate the area around the anode (especially where carbon-based anodes are used, where the carbon also turns into CO_2) and can lead to etching of the concrete surface and attack on the cement paste and even some aggregates once the alkalinity is consumed.

We can therefore see that three factors must be taken into account when controlling our CP system:

1. There must be sufficient current to overwhelm the anodic reactions and stop or severely reduce the corrosion rate.
2. The current must stay as low as possible to minimise the acidification around the anode and the attack of the anode for those that are consumed by the anodic reactions.
3. The steel should not exceed the hydrogen evolution potential, especially for pre-stressed steel to avoid hydrogen embrittlement.

4.5 GALVANIC CATHODIC PROTECTION MECHANISM

The explanations so far have described CP assuming that the current comes from a power supply and is delivered through an impressed current anode into the concrete. However, there are also galvanic anodes that can be installed to control corrosion.

These anodes rely on the fact that there is an electrochemical series of metals with one of the most noble being gold, which does not corrode, and one of the least noble being sodium, which combusts explosively and spontaneously on contact with air. When a less electronegative metal (more positive potential, more noble) is in electrical contact with a more electronegative metal (more negative potential, more likely to corrode), it is the same as having anodes and cathodes on the same piece of metal as shown in Figure 4.1. If there is moisture between them and oxygen available, the potential difference will lead to a current flow. This is the galvanic effect and if the potential difference and resistance in the cell allow sufficient current to flow and for the potential on the less noble anode can move the potential on the more noble steel, then the steel will be cathodically protected.

Zinc is a widely used galvanic anode for steel. In sea water, it has a potential (voltage) of about −1.0 V against a standard copper/copper sulphate reference cell. Passive steel reinforcement in concrete has a potential of about −0.10 V, while actively corroding steel reinforcement in concrete usually has a potential −0.35 to −0.50 V.

Based on the potential difference between the reinforcement steel and the anode, we can see that a galvanic zinc anode for reinforced concrete CP has between 0.5 and 0.9 V to drive the CP current to the steel. The driving voltage of the galvanic anode is determined by the type of material used. Only modest increases in the driving voltage are possible by changing the material, for example, an increase of 0.5 V is possible when using magnesium as the galvanic anode.

As we cannot significantly increase the galvanic anode driving voltage, and by using Ohm's law, $V = IR$ we can see that the electrolyte (concrete) resistivity R will have a direct impact on the amount of CP current I we can deliver to the steel for a given voltage V.

4.6 CATHODIC PROTECTION CURRENT DENSITIES

When setting the current on an impressed current CP system, it must be high enough to control the corrosion but low enough to minimise the deleterious effects on the anode and maximise its life.

Accepted cathodic prevention current density levels are 0.2–2.0 mA/m^2 based on the total surface area of the steel affected by the application of the protection current. As this is easily achieved using any of the normal anode systems, the distribution of the current becomes the most important design aspect. The anode system must be able to provide a reasonable distribution of the current to the steel. Wide spacing of the anode should be avoided even though the anodes are capable of the design output. For normal steel mats, a maximum anode spacing of 400 mm would be recommended. Further information is given in Chapters 5 through 7 on design.

However, while we design a system with maximum output current based on the steel surface area and the expected conditions, we can only adjust the current by measuring the potentials against embedded reference electrodes and ensuring that they stay within the required control criteria.

4.7 CATHODIC PROTECTION POTENTIALS

As shown in Figure 4.1, we cannot measure the potential of the anodes and cathodes on the steel. We can only see a mixed potential. This is further complicated when we apply CP.

Figure 4.2 shows a simple schematic of a reference electrode embedded in concrete between an impressed current anode and a reinforcing bar. When the CP current flows between the anode and the steel and a reference electrode is positioned between them, this CP current flow will also be picked up by the monitoring circuit, making the steel potential appear more negative than it actually is.

In very simple terms, if the CP system voltage is 10 V and the reference electrode is half way between the anode and steel in a uniform resistance concrete, the apparent potential between reference electrode and steel will be 5 V. Only if we turn off the current can we measure the actual potential between the reference electrode and the steel. However, as soon as we do, the steel will depolarise and potential will change.

It is therefore essential to measure the 'instant off' potential between the reference electrode and the steel rather than the 'on' potential. This

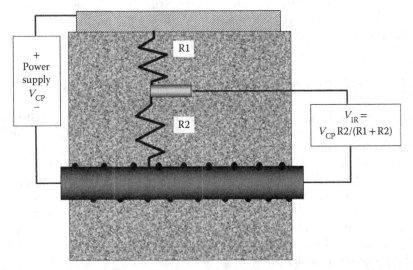

Figure 4.2 The 'IR' drop effect on a reference electrode potential measurement caused by an impressed current flow.

is measured between 0.1 and 1 seconds after switching off. Although in electrochemical terms the instant off can be measured a few microseconds after switch off, in practice, on a large zone of steel and anode capacitive effects and current flows round the systems mean that it is often difficult if not inaccurate to measure too quickly after switch off.

We will always be referring to instant off potentials when discussing measurements on CP systems.

The potential measurements are therefore a measure of how effectively the current is controlling corrosion. We can therefore try to keep the potentials within a given range or we can look at how the potentials decay after switch off to see how much the system has polarised and, therefore, how likely that we have suppressed all anodes on the steel.

4.8 CONTROL CRITERIA

The normal way to control a CP system is to measure the potentials of the steel against embedded reference electrodes and adjust either the current output or the voltage from the power supply.

There are two major types of control criteria, absolute (instant off) potentials and potential decay or depolarisation criteria. There are other, less well-used criteria including null probes and potential versus current plots.

4.8.1 Absolute potentials, maxima and minima

The first absolute potential criterion is the hydrogen evolution limit. This is generally set at −1100 mV versus all embedded silver/silver chloride/0.5 M potassium chloride (Ag/AgCl/0.5 M KCl) reference electrodes. For pre-stressing steel, it is set at a less negative, more conservative potential of −900 mV versus Ag/AgCl/0.5 M KCl.

The higher limit to achieve corrosion control is an absolute instant off potential of −720 mV versus Ag/AgCl/0.5 M KCl. This is a criterion that was developed for buried and submerged steel pipes and is widely used throughout the CP industry. The problem is that potentials have to be controlled between −720 and −1100 mV, which may be impossible for concrete due to its high resistance.

4.8.2 Potential shifts, 100 and 150 mV

The option that is most widely used in concrete and also increasingly used for CP of buried or submerged steel structures is to use potential shift criteria. This reflects the requirement to polarise the cathodic areas to those of the anodic areas so that no corrosion currents can flow between them.

For passive steel in atmospherically exposed concrete (i.e. with natural potentials less negative than −0.150 mV with respect to Ag/AgCl/KCl), any negative shift in potential will provide additional corrosion protection; however, this is not definitive. Current recommendations include seeking a 50-mV negative shift with the application of cathodic prevention (see Chapter 8); this ensures that corrosion cannot initiate on new structures or older structures where corrosion has not yet started but is vulnerable to corrosion. This potential shift would appear to demonstrate sufficient applied external current.

For situations where the steel is corroding, that is, in chloride-contaminated, carbonated or damaged concrete, a shift of 100 mV is generally accepted as being sufficient to mitigate naturally occurring corrosion currents. There is a comprehensive review of the 100-mV criterion published by NACE (2008).

In the first case, the shift in potential can be measured as the change from the 'natural' or 'rest' potentials, before current is applied, to an 'instant off' potential with the system running. However, as explained in Section 4.1, all sorts of benign reactions will occur on the steel surface, thus changing the local chemical environment at the steel surface. After a few weeks or months, the 'natural' or 'rest' potentials would be changed. It should be possible to switch the system off for several weeks until potentials stabilise and then reapply the shift criterion from rest. However, to ensure that the system is providing protection and that measurements and adjustment are untaken efficiently, it is normal to measure the potential decay or depolarisation from 'instant off' to 'off' over a convenient period.

Both the magnitude of the depolarisation and the length of the depolarisation are subject to discussion, dispute and interpretation.

Bennett and Mitchell (1989) concluded that 150 mV of polarisation was needed to achieve corrosion control in the most aggressive conditions. This was partly based on the fact that the Tafel slope (the gradient of potential vs. log of current) may be as high as 150 mV per decade for steel in concrete and, therefore, 150 mV is required to reduce the corrosion rate by 90%. Bennett and Broomfield (1997) confirmed this by reviewing work showing the potentials and current densities needed to reduce the corrosion rate to 2.5 µm/year, the common criterion used for effective CP. At >1.6 wt% chloride, or 6 kg/m³, this was 150-mV cathodic polarisation or 17 mA/m².

4.8.3 E-Log i tests

An empirical test method (E-Log i) has been developed to determine the steel potential versus applied current relationships in particular cases. As varying levels of current is applied to the steel, the potential E_m can be measured and recorded.

From this data, an E-Log i curve can be developed. Over a section of the curve close to its natural potential, there should be an almost linear portion from where the relationship between applied current and resulting potential shift can be determined. There should be a change in gradient as the steel moves from anodic to cathodic conditions.

The limitations of this technique are that it is limited to first polarisation only and it requires orders of magnitude of applied current. It is therefore quite difficult to determine the start and end points of the test. It is also difficult to arrange the testing equipment on sites and there remains the possibility of overprotection that can damage the concrete. It is not generally considered to be a practical on-site test. However, some U.S. companies have the equipment and still use it.

The protection current required is taken to be the current at the start of the linear portion of the E-Log i curve, which is sometimes difficult to achieve or detect in real cases.

4.9 CRITERIA FOR PROTECTION IN THE MAJOR STANDARDS

As stated in Section 4.8, there are useful ways of measuring potentials of steel against embedded reference electrodes to ensure that the current is sufficient to control corrosion without causing excessive damage to the concrete around the anode.

4.9.1 NACE standard practices 0290-2000, SP 0100:2008, 0408-2008

The first major standard on CP of steel in concrete was NACE RP 0290, now revised as NACE SP 0290-2000. This standard is for CP of atmospherically exposed reinforced concrete. The standard was initially developed when the major applications of CP in North America were to bridge decks. It contains two criteria, a 100-mV polarisation decay or development and the E-log i test.

The 100-mV criterion is described in some detail and the two example curves are given of polarisation development from rest potential and depolarisation from instant off. There is a requirement that '100 mV of polarisation should be achieved at the most anodic location, typically in every 46 m^2 area or zone or at artificially constructed sites'.

The observant reader will note that the discussion in Sections 4.1 and 4.8 refers to moving the cathodic areas below the potential of the anodic areas. NACE requires the most anodic areas to shift by 100 mV.

There is also a note of the fact that environmental conditions can change over the polarisation or depolarisation measurement period and that the engineer may need to take account of this. It also notes that oxygen availability can be reduced by water saturation, so longer decay periods may be required for less permeable, coated or water-saturated structures.

NACE SP 0290 also refers to the fact that no polarisation is required if the steel is passive, that is, the potential is less than −200 mV versus copper/saturated copper sulphate (−150 mV Ag/AgCl/0.5 M KCl).

There is a brief paragraph on the E Log i test, saying that it can be used to determine the initial CP current.

NACE SP 100-2008 for concrete pressure pipes and mortar-coated pipes uses exactly the 100-mV polarisation/depolarisation criteria as SP 0290 but with a single, more complex plot of the polarisation decay and development curves.

NACE SP 0408-2008 is for buried and submerged reinforced concrete. It has the same 100-mV decay/development criteria with the more sophisticated plot of potential versus time as SP 100. In addition, there is an absolute potential criterion −850 mV versus copper/saturated copper sulphate (−800 mV Ag/AgCl/0.5 M KCl).

There is also an additional limit of no more than −1000 mV versus copper/saturated copper sulphate (−950 mV Ag/AgCl/0.5 M KCl) for 'high strength steel (>69 MPa) … susceptible to hydrogen embrittlement'.

4.9.2 BS EN ISO 12696:2012

The newly published ISO standard was developed from the European standard BS EN ISO 12696:2012. It covers reinforced concrete that is atmospherically exposed, buried or submerged. It also covers CP by galvanic anodes as well as impressed current.

The criteria start by giving absolute potential limits of an instant off potential of −1100 mV versus Ag/AgCl/0.5 M KCl for plain reinforcing steel and −900 mV for pre-stressing steel. There are then three options and the CP zone should meet any one of the three (not all three as suggested by some engineers and owners of structures).

The first is an instant off potential of −720 mV versus Ag/AgCl/0.5 M KCl, that is, the criterion for buried or submerged steel. The second is a potential decay over a maximum of 24 hours of 100 mV from instant off. The third is a decay of 150 mV over a period longer than 24 hours.

There is a series of informative notes regarding the stability of the readings in changing environments, the slowing of depolarisation in situations where oxygen availability is limited by saturations or coatings as seen in the NACE criteria.

There is a reference to how to deal with galvanic anode systems that cannot be adjusted. If none of the criteria are met and it is not possible to increase the current (e.g. by adding more anodes), a risk assessment is required.

4.10 CONCLUDING REMARKS

This chapter has summarised the present understanding of the mechanism of CP and the practical application of control criteria. The NACE standards and the ISO standard are quite similar, giving absolute potential limits and potential shift criteria. The values of the absolute potentials vary slightly and the methods of carrying out measurements are also slightly different.

The major practical difference is that the most widely used 100-mV shift criterion has no time limit in the NACE standards, although reference is made to the slowing of depolarisation in saturated concrete. The ISO standard has 100 mV for up to 24 hours and 150 mV thereafter to reflect the 150-mV total depolarisation suggested by the literature such as Bennett and Broomfield (1997).

Both sets of standards acknowledge the need for conditions to be stable enough for conditions to be unchanged at the steel surface that the reference electrode is measuring over the period of measurement.

REFERENCES

ASTM C876:09. *Standard Test Method for Corrosion Potentials of Uncoated Reinforcing Steel in Concrete*, American Society for Testing and Materials, West Conshohocken, PA.

Bennett, J.E. and Broomfield, J.P. (1997). Analysis of studies of cathodic protection criteria for steel in concrete, *Materials Performance* 36(12): 16–21.

Bennett, J.E. and Mitchell, T.A. (1989). Depolarization testing of cathodically protected reinforced steel in concrete, Proc NACE Corrosion, NACE International, Houston, TX.

Bentur A., Diamond, S., and Berke, N.S. (1997). *Steel Corrosion in Concrete: Fundamentals and Civil Engineering Practice*, E & FN Spon, London, UK.

Broomfield, J.P. (2007). *Corrosion of Steel in Concrete: Understanding Investigation and Repair*, 2nd Edition, Taylor and Francis, London, UK.

BS EN ISO 12696:2012. *Cathodic Protection of Steel in Concrete (ISO 12696:2012)*, British Standards Institute, London, UK.

NACE. (2002). *State of the Art Report: Criteria for Cathodic Protection of Prestressed Concrete Structures*, Report No. 01102, NACE International, Houston, TX.

NACE. (2008). *One Hundred Millivolt (mV) Cathodic Polarization Criterion*, Report No 35108, NACE International, Houston, TX.

NACE SP 0290. (2007). *Impressed Current Cathodic Protection of Reinforcing Steel in Atmospherically Exposed Concrete Structures*, NACE International, Houston, TX.

NACE SP 100. (2008). *Cathodic Protection to Control External Corrosion or Concrete Pressure Pipelines and Mortar Coated Steel Pipelines for Water or Waste Water Services*, NACE International, Houston, TX.

NACE SP 0408. (2008). *Cathodic Protection of Reinforcing Steel in Buried or Submerged Concrete Structures*, NACE International, Houston, TX.

Chapter 5

History and principles of cathodic protection for reinforced concrete

Paul M. Chess and John Broomfield

CONTENTS

5.1 HISTORY OF CATHODIC PROTECTION

Sir Humphry Davy is credited with first developing cathodic protection (CP) in 1824. He discovered that iron would protect copper plating on ship hulls from corrosion. However, the corrosion of the copper was what stopped the marine fouling of the hulls, so this first experiment in galvanic CP was abandoned.

Thomas Edison is credited with the first application of impressed current cathodic protection (ICCP), again to ship hulls in the 1890s. In the 1920s, CP was applied to steel gas pipes in North America and from there the CP industry expanded to cover a wide range of buried and submerged steel structures using impressed current and galvanic anodes.

5.2 FIRST SYSTEMS FOR REINFORCED CONCRETE BRIDGES

The first major steps to electrochemical corrosion control treatment of reinforced concrete occurred in the United States as early as 1959 [1] when Richard Stratfull applied a trial ICCP to a bridge deck and substructure suffering from chloride attack due to de-icing salt ingress. He then went on to develop a CP system for bridge decks, which was applied to two bridges in California reported in 1974. These were still working in 1989 [2].

Between 1973 and 1989, a total of 287 systems were installed on U.S. interstate highway bridges, predominantly on structures suffering from de-icing salt attack due to the lack of any surface protection or waterproofing on the deck [2]. Many more systems were applied to other structures as well as to other bridges owned by the states, counties and cities in the United States and Canada.

The first trials and full-scale ICCP systems in the United Kingdom and Australia were undertaken in the mid to late 1980s [3]. These were done on buildings suffering from the deliberate addition of calcium chloride as a set accelerator in the United Kingdom and on jetties due to marine exposure and cement works due to sea salt exposure in Australia. Since then, over 2 million m² of ICCP has been applied to reinforced concrete structures worldwide.

Galvanic anodes were first developed in the late 1990s, initially for marine exposed substructures and then for local protection around patch repairs. Many millions of galvanic anodes have now been sold worldwide.

5.3 DEVELOPMENT OF CATHODIC PROTECTION ANODES FOR STEEL IN CONCRETE

The first CP anodes were developed from the standard silicon iron anode in a coke breeze backfill, used to protect buried pipelines. Stratfull took the anode, had it made as a flat 'pancake' and surrounded it in a coke breeze–loaded asphalt. This anode was used from Strafull's first installations in 1974 until the 1990s and beyond. Its chief drawback was that it added to the load on the deck, raised the deck level and, being highly permeable to water, could give freeze thaw problems on decks with inadequate air entrainment (which controls freeze thaw damage).

The first major step forward in anode design was the development of the mixed metal oxide–coated titanium mesh in the 1990s. This anode had been developed for the chloralkali industry to work at very high current densities to generate chlorine gas from brine. It could pass very high currents and was extremely durable. The inventor, Jack Bennett of Eltech, made it into a fine expanded mesh, applied it to bridge decks and placed a layer of concrete over it, as shown in Figure 5.1. After its success on decks, it was also applied to substructures with a suitable sprayed repair mortar applied.

While North America concentrated on problems on their bare, unprotected bridge decks, in the United Kingdom, there was interest in applying CP to buildings suffering from cast in calcium chloride, marine-exposed structures and bridge substructures. In Europe, there was a policy of applying waterproofing membranes to bridge decks to the de-icing salts concentrated on the beams and columns underneath expansion joints causing corrosion problems below the deck.

Figure 5.1 Early application of a mesh and overlay anode system on a U.S. bridge deck.

Figure 5.2 A conductive coating anode applied in an industrial plant in Australia in the 1990s.

Following initial trials in the United States, conductive coatings were trialled and then used in earnest in the United Kingdom and Australia in the mid-1980s, along with Hong Kong and other Far Eastern countries. An example is shown in Figure 5.2.

In Italy, the first 'cathodic prevention' systems were installed on new bridge decks during construction using mixed metal oxide–coated titanium mesh ribbons [4]. This led to the development of a major market for cathodic prevention systems in the Arabian Gulf region where the level of chloride contamination combines with the drying conditions evaporating moisture to concentrate chlorides and give a highly corrosive environment.

The mixed metal oxide–coated titanium anode is now available in a number of configurations, including the expanded mesh, ribbon and probe anodes embedded din holes drilled in concrete. Figure 5.3 shows some examples.

The first experiments with galvanic CP for concrete were done by Florida Department of Transportation in the 1990s [5]. These were applied to columns in the sea on bridges linking the islands of the Florida Keys. These included a jacket system of a permanent form enclosing an expanded zinc mesh and a porous grout, as shown in Figure 5.4.

Florida Department of Transportation also used thermal-sprayed zinc as a galvanic anode (see Figure 5.5), which had been developed as an impressed current anode by California Department of Transportation [6]. Thermal-sprayed zinc was further developed as a galvanic anode, first by the application of a humectant that kept up the moisture level to increase current flow in non-marine applications. This was followed by the development of an aluminium–zinc–indium alloy with a higher current output than pure zinc.

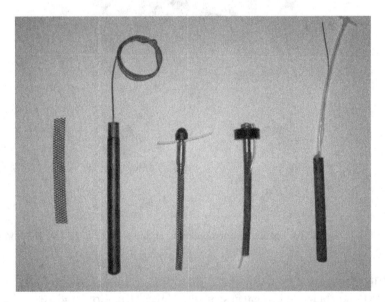

Figure 5.3 Embeddable anodes: Ribbon anode use in CP for new construction fixed to the reinforcement by insulating spacers or embedded in slots in the concrete on older corroding structures (left side). Three types of embeddable, mixed metal oxide–coated titanium anodes are shown in the centre. On the right side is an embeddable conductive ceramic anode.

Figure 5.4 A jacketed zinc galvanic anode system applied to columns on a jetty in the Channel Islands, United Kingdom.

Figure 5.5 Thermal-sprayed zinc being applied to a bridge substructure in the Florida Keys in the 1990s.

In the United Kingdom, researchers at Aston University came up with an embeddable galvanic anode that used a lithium-rich mortar around a zinc anode to prevent the passivation of the anode, which occurs when zinc is embedded in normal mortar or concrete. Initially designed to prevent the 'incipient anode' or 'ring anode' effect where corrosion initiated

Figure 5.6 A galvanic anode in a patch repair to provide local protection around the repair and prevent the 'incipient anode' effect of corrosion around a repair.

round patch repairs to be embedded in repairs, as shown in Figure 5.6. These anodes could not be considered a proper CP system anode. They were then developed to be installed on a grid to provide more extensive protection.

Further developments of embeddable galvanic anodes came with the development of an anode with a humectant admixture to activate it rather than the lithium. This has been followed by a 'hybrid' embeddable anode system that is first energised like an impressed current anode, attracting chlorides towards it and acidifying the surface to activate it so that it can then function as a galvanic anode.

Another surface-applied galvanic anode is the adhesive zinc sheet. This uses a conductive hydrogel adhesive, as shown in Figure 5.7.

The choice of anode is fundamental to the design of a CP system. While a galvanic anode may be simpler to install and require no power supply and minimal wiring, there is no control, no assurance that the system is working and a limited life of anodes, typically 10–20 years. Anodes tend to be intrusive either surface-mounted or embedded in large holes in the concrete.

Impressed current systems are more complex and therefore higher first cost with an ongoing maintenance cost. However, they can be designed with anode lives of over 100 years. The range of anodes means that they can be suitable even for historic listed structures.

Figure 5.7 Zinc hydrogel adhesive galvanic anode applied to a balcony fascia before the application of a protective/sealing coat.

5.4 CURRENT STANDARDS FOR CATHODIC PROTECTION

The first widely used standard for ICCP was the NACE International Recommended Practice RP 290. This was first published in 1990. It was followed by the European Standard EN ISO 12696 that was published in 2000, at which time RP 290 was revised. These covered the application of ICCP to atmospherically exposed steel in concrete.

Since then, NACE has published two test methods for impressed current anodes, TM 0294 for embeddable anodes (mainly aimed at mixed metal oxide–coated titanium anodes) and TM 0105 for organic-based conductive coating anodes. NACE has also published standard practices on CP of concrete and concrete-coated pipes (NACE SP 0100) and a standard on CP of buried and submerged concrete structures (NACE SP 0408). SP 0290 was last revised in 2007 and is due for revision at the time of going to press.

The European Standards Organisation (CEN) has published a series of standards on CP since 2000. One that is relevant to the practice of CP of steel in concrete is BS EN 15257 on competence levels and certification of CP personnel. This has led to the setting up of training courses in European countries to provide certification at the basic operative, supervisor and designer level specifically for CP of steel in concrete.

CEN has also produced a two-part standard on electrochemical realkalisation and on chloride extraction, CEN/TS 14039 parts 1 and 2. NACE has a single equivalent standard, SP 0107.

Most recently, the European Standard on CP has been revised to cover all aspects of CP of steel in concrete, impressed current, galvanic, atmospherically exposed, buried and submerged. This has been adopted by the International Standards Organisation and is therefore designated BS EN ISO 12696:2012 [7].

REFERENCES

1. Stratfull, R.F. 'Progress Report on Inhibiting the Corrosion of Steel in a Reinforced Concrete Bridge'. *Corrosion* 15, no. 6 (June 1959): 331t–334t.
2. Broomfield, J.P., and J.S. Tinnea. *Cathodic Protection Reinforced Concrete Bridge Components*. SHRP Report SHRP-C/UWP-92-618 (1992). http://onlinepubs.trb.org/onlinepubs/shrp/SHRP-92-618.pdf.
3. Broomfield, J.P., P.E. McAnoy, and R. Langford. 'Cathodic Protection for Reinforced Concrete: Its Application to Buildings and Marine Structures'. *Paper presented at the Corrosion of Metals in Concrete*, San Francisco, CA (1987).
4. Bertolini, L., F. Bolzoni, L. Lazzari, and P. Pedeferri. 'Applications of Cathodic Protection to Steel in Concrete'. *International Journal for Restoration of Buildings and Monuments* 6, no. 6 (2000): 655–668.
5. Kessler, R.J., R.G. Powers, and I.R. Lasa. 'Update on Sacrificial Anode Cathodic Protection on Steel Reinforced Concrete Structures in Seawater'. *Corrosion* 95, no. 516 (1995).
6. Apostolos, J.A., D.M. Parks, and R.A. Carello. 'Cathodic Protection Using Metallized Zinc'. *Materials Performance* (December 1987): 22–28.
7. BS EN ISO 12696:2012. *Cathodic Protection of Steel in Concrete (ISO 12696:2012)*, British Standards Institute, London, UK.

Chapter 6

Immersed cathodic protection design

Arnaud Meillier

CONTENTS

6.1 INTRODUCTION

Most of this work was conducted during the author's employment with the Materials and Corrosion department of Mott Macdonald in Altincham, United Kingdom. The author wishes to particularly thank Professor Paul Lambert and Dr Chris Atkins for their support and inputs. Reinforcing steel embedded in concrete, by definition, has a very low susceptibility to corrosion as the alkaline nature of the cement paste provides favourable conditions for the build-up and maintenance of a stable oxide film on the steel surface. However, the passive nature of reinforced concrete (RC) may be modified by the migration of aggressive species from the external environment to which the structure is exposed. The mechanisms of corrosion for atmospherically exposed RC structures are well documented, and cathodic protection (CP) has been proven to be a technically and economically effective method for stopping and preventing reinforcing steel corrosion. This particular application of CP has been employed for a large number of civil structures over the past 20–30 years and is now at a stage where the installation, design and performance criteria of such systems are largely mastered and adequately documented.

Conversely, due to the limited number of applications to date, there is considerably less experience in the CP of RC structures in earth or water. The application of CP in such cases requires skills in 'traditional' CP technology (pipelines, tanks, jetties etc.) coupled with a comprehensive understanding of the electrochemical behaviour of steel embedded in concrete.

The following text focuses mainly on the procedure of CP design for buried or submerged RC structures, combined with a consideration of the mechanisms of corrosion and the influence of the external environment.

6.2 BURIED CONCRETE STRUCTURES

6.2.1 General

As with its above-ground counterpart, RC structures buried in the earth may be subject to corrosion under a number of particular conditions. However, the oxygen availability being markedly reduced for buried structures, the rate of attack may be significantly lower. It must also be acknowledged that the difficulties in conducting inspections of buried structures mean than many concrete defects go undiscovered unless they result in damage or failure.

The design and construction of buried civil engineering structures are generally preceded by geotechnical surveys that should provide sufficient physico-chemical data to define the soil mechanics and geological conditions at a site. These soil characteristics, supported by the relevant civil engineering codes of practices or standards, may influence the concrete mix specification and the need for any passive protection against external aggressive species such as tanking. Such preventive measures adopted at the design stage are considered to belong to the civil engineering discipline and are out of the scope of this chapter. Nevertheless, a brief description of the likely corrosion mechanisms is given in the following discussion.

CP of mortar-coated steel pipelines and concrete cylinder pipes (CCP) is probably the most widespread application of CP for buried RC structures with the first such application being as early as 1946 in Algeria [1].

Over the past decade or so, the increasing awareness of the long-term economic advantage of installing CP systems in RC structures exposed to aggressive environment at the time of construction ('cathodic prevention') has lead to a number of applications where the protection of buried or submerged reinforcing steel was included in the CP scheme.

However, it should be noted that in the latter case, the type and placement of the anodes is, in most cases, similar to that encountered with atmospherically exposed concrete (i.e. anodes placed within the structure itself) and therefore significantly differ with the overall topic of this chapter.

6.2.2 General consideration with respect to corrosion

6.2.2.1 Above-ground concrete corrosion mechanisms

For atmospherically exposed concrete structures, the disruption of the passive layer on the reinforcing steel is predominantly caused by the ingress of two very different species from the surrounding environment:

1. Chloride ions, once at the steel—concrete interface in the presence of both moisture and oxygen, may cause localised corrosion or pitting of the steel. This type of corrosion is often referred to chloride-induced corrosion, which mainly influences the anodic polarisation curve in the corrosion cell as illustrated in the simplified Evans diagram (E-Log i diagram) in Figure 6.1.
2. Carbon dioxide, as present in the atmosphere, can cause the reinforcing steel to corrode by means of a different mechanism. The carbon dioxide gas dissolves into the pore water present within the concrete to give carbonic acid (H_2CO_3). This then neutralises the alkalis in the pore water to form calcium carbonate ($CaCO_3$) resulting in a loss of alkalinity at the steel—concrete interface and thus allowing corrosion to occur. This type of corrosion process is known as carbonation, which causes the steel to corrode uniformly. This type of corrosion also influences the anodic polarisation curve in the corrosion cell as illustrated in Figure 6.2.

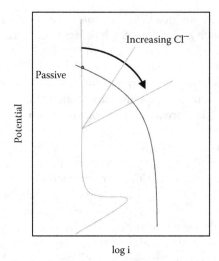

Figure 6.1 Chloride-induced corrosion schematic Evans diagram.

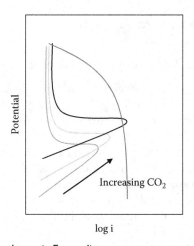

Figure 6.2 Carbonation schematic Evans diagram.

Of these two mechanisms, chloride-induced corrosion is considered to be the most significant with respect to above-ground concrete structures. This is also true for buried and submerged structures where exclusion of the atmosphere makes carbonation unlikely. Based on the oxygen availability requirement for corrosion to take place, being the predominant cathodic reactant, it is often assumed that reduced oxygen availability results in a low risk and rate of corrosion. Careful consideration must be given when adopting such reasoning as it has been reported by several investigators that reinforcement corrosion is thermodynamically feasible under very low oxygen conditions.

6.2.2.2 Buried concrete corrosion mechanisms

6.2.2.2.1 General

RC buried in earth may also be subject to chloride-induced corrosion if oxygen is available. As discussed in Section 6.2.1, the design of any buried concrete structure is typically preceded by a geotechnical survey that may influence the construction specification. Among the soil data surveyed are the presence of groundwater and its composition and the sulphate content of the soil or water. Sulphates are considered to be aggressive to ordinary Portland cement (OPC)—based RC as they can react with the tricalcium aluminate hydrate component of the cement paste that may cause softening and breakdown of the concrete. The latter may induce significant loss in structural capacity due to the loss of concrete cover, and once the reinforcement is exposed to the soil, corrosion may proceed at a greater pace. It may also lead to preferential corrosion of the steel in contact with the soil due to the galvanic couple created with the embedded rebars, plus the associated small anode—large cathode surface area ratio. Depending on the concentration of the sulphate species in the soil or groundwater, the civil designer is equipped with various options to ensure the concrete durability by, for example, increasing the concrete cover, modifying the concrete mix characteristics or applying a tanking system onto the concrete surface [2,3]. In extreme cases, concrete at risk of sulphate attack may require the use of sulphate-resisting Portland cement (SRPC) with a maximum free water/cement ratio of 0.45 in combination of a protective coating [2,3].

It should be noted that SRPC has a lower chloride ion—binding capacity than OPC, which implies that the use of SRPC without a coating could render the situation worse in terms of chloride-induced corrosion.

In an evaluation addressing the reinforcement corrosion in a stagnant saline environment, Morgan [4] suggested a corrosion mechanism due to the sulphate ions similar to that for chlorides and resulting pitting corrosion. This may also be true for soil. However, the investigator suggested that this mechanism would not be as important as chloride-induced corrosion.

Another corrosion parameter to take into account is that any buried metallic structure is potentially subject to stray current corrosion caused by neighbouring DC traction or third-party CP systems, which could cause accelerated corrosion of the reinforcement. One valid mitigation method is to provide the affected structure with a dedicated CP system, which would help in counteracting the detrimental electrochemical potential anodic shift. By shifting the potential more electro-negative by the application of CP, the induced anodic shift by the interfering structures no longer induces corrosion but a reduction in the protective potential. This is one of the reasons why many CCP and pre-stressed pipelines are fitted with CP.

Techniques to determine site aggressivity towards buried concrete structures are described by a number of authors, nominally Cherry [5] presented comprehensive information with regard to corrosion of buried RC structures and Benedict [6] suggested a flow diagram for corrosion protection on CCP.

It should be emphasised that a clear distinction must be made between general soil characteristics and aggressivity for temperate countries and the particular case of the countries neighbouring the Arabian Gulf seaboard (GCC* countries). In addition to the high relative humidity and temperature of the climate, the geology is often characterised by the presence of saline groundwater and Sabkha. Care must be exercised in applying western civil engineering practice in these countries and specific considerations must be given to adapt construction practices to the higher corrosive nature of these soils [7]. Barrier systems such as waterproofing membranes or tanking have been widely used in the Middle East to prevent the ingress of water and chlorides from the ground to foundations although it has been reported that they were ineffective for total, long-term concrete durability protection [8,9].

6.2.2.2.2 Emphasis on key soil-related parameters

6.2.2.2.2.1 SOIL ELECTRICAL RESISTIVITY

Similar to buried steel structures, RC structures in low electrical resistivity soil are generally more susceptible to corrosion; however, the presence of highly aerated soil conditions must be not disregarded. Buried RC structures are candidates for an additional corrosion mechanism not suffered by concrete exposed to the atmosphere. Depending on the soil conductivity, the corrosion current is no longer restricted by the electrical resistance of the concrete as an electrolytic path is then possible through the soil.

Traditionally, soil resistivity measurements were used on buried steel structures to assess the soil corrosivity. Due to its simplicity, the technique is conveniently employed in the field but only provides a preliminary indication of the probability of corrosion. Conclusions made with regard to the corrosion risk of a soil based solely on conductivity would not be adequate as other parameters such as chloride and sulphate ions, moisture and pH are generally needed to complete the assessment. Comprehensive appraisal of soil corrosivity is generally most satisfactorily conducted by the use of the 'global index method' [10]. The method consists of measuring a number of selected soil parameters with each one awarded a weighting factor. The summation of the factors determines the corrosivity of the soil. An example of this method applicable in the United Kingdom would be the ranking

---

* GCC: Gulf Cooperation Council, which comprises Kingdom of Saudi Arabia, Kuwait, Kingdom of Bahrain, State of Qatar, the United Arab Emirates (UAE) and the Sultanate of Oman. Founded on 26 May 1981, the aim of this collective is to promote coordination between member states in all fields to achieve unity.

system for soil corrosivity utilised by the Ministry of Defence to determine protection measures for steel fuel pipelines [11].

However low resistivity soils, in most instances, are often associated with the presence of species aggressive for both reinforcement and steel (chloride and sulphate) and moisture/water. In the case of buried RC, it may be anticipated that the rate of attack is less than for buried bare or coated steel owing to the corrosion resistance conferred by the concrete.

In an early 1960s evaluation on corrosion and CP of buried pre-stressed concrete structures, Heuzé [12] reported European industry practices where CP was not generally implemented for soil resistivities over 1000 Ω-cm although in exceptional circumstances CP may have been used for soil resisitivities between 1000 and 2000 Ω-cm. Pipe and tank manufacturers are reported to consider the real danger of corrosion to be at resistivities less than 500 Ω-cm. There were a small number of cases in the 1950s where CP was applied to buried pre-stressed concrete pipeline and tank for soil resistivities greater than 6000 Ω-cm. However, it should be noted that all the preceding applications took account of the mortar and pre-stressed steel quality available at the time. A publication from 1986 describing the recommended practices in France [13] quoted that implementation of CP would be beneficial for soil resistivity less than 6000 Ω-cm for the majority of RC structures with the concrete quality most often encountered. It also indicated that CP would be a relevant precautionary measure for inferior concrete quality buried in earth at resistivities of less than 5000 Ω-cm.

Based on Australian practical experience, Gourlet and Moresco [14] suggested the following soil conditions where CP could be implemented on pre-stressed concrete cylinder pipelines (PCCP):

* Less than 200 Ω-cm: CP is essential
* 2000–3000 Ω-cm: CP is recommended
* 3,000–10,000 Ω-cm: CP is normally required
* Over 10,000 Ω-cm: CP is not necessary

The investigators also stated that CP is recommended for soils that are very corrosive to concrete or conductive to rapid groundwater movement.

In a comprehensive assessment of the corrosion and protection of CCP, Benedict [6] proposed guidelines for when to bond electrically discontinuous pipe segments based on the various surveyed soil characteristic data. Bonding is generally required when the corrosion risk of the soil is evaluated to be significant and allows remediation by CP or the application further monitoring measures with a view to ascertaining the actual risk of corrosion. Among several prerequisites for bonding are presented conditions where the soil resistivity is less than 1300 Ω-cm and chloride concentrations greater than 500 ppm, but Benedict also included cases with soil containing sand (which are in most instances highly permeable) subject to numerous wet/dry cycles (e.g. varying groundwater table) and a substantial chloride content at pipe depth. Similarly, in a document produced by a North American association [15], it is suggested that a soil could be considered aggressive to CCP when it exhibits resistivity readings below 1500 Ω-cm with water-soluble chloride content greater than 400 ppm.

6.2.2.2.2.2 ELECTRICAL RESISTIVITY OF CONCRETE IN SOIL

Resistivity for atmospherically exposed concrete structures generally ranges from 10,000 to 100,000 Ω-cm depending on exposure condition, chloride and moisture content [16]. Foundation metalwork in concrete is sometimes used as an effective earth electrode as indicated in BS EN 7430:1998 [17], and it is suggested that the electrical resistivity of concrete when buried in soil, except in dry conditions, could be expected to be about 3000–9000 Ω-cm. The reason for this is that the concrete is considered hygroscopic and once the concrete is buried its moisture content will reach an equilibrium state with that of the soil. It should be emphasised that the preceding quoted values originated from British electrical earthing experience and these values reflect soil characteristics of that geographical region; therefore, care must be exercised when using these guidelines outside temperate climates. It was also reported that concrete resistivities ranged from 3000 to 5000 Ω-cm for RC rod used as grounding electrodes in soil [18]. This change of concrete resistivity was also reported by Franquin [1] who indicated a value of 8000 Ω-cm for CCP after burial. Gourlet [19] reported a reading of 10,000 Ω-cm for a similar application. Heuzé [20] indicated a conservative value of 5000 Ω-cm for buried CCP in soil with a resistivity of 1000–2000 Ω-cm involving CP. Benedict [6] recommended, possibly for reasons similar to those expressed in the preceding discussion, that any potential survey on recently laid and buried CCP should be carried out after 6–12 months after pipeline burial to allow the CCP to reach a state of equilibrium with the soil environment.

6.2.2.2.2.3 ELECTROCHEMICAL CORROSION (NATIVE) POTENTIAL

For most buried and submerged metallic structures, the measurement of the electrochemical potential can assist in the assessment of the corrosion activity. This simple and convenient tool can be used for existing below-ground concrete structures, but if it is used as a stand-alone technique, it may result in rather complex interpretations and may not give conclusive results regarding the corrosion appraisal.

For above-ground concrete structures, the use of half-cell potential mapping (with a reference electrode placed onto the concrete cover) continues extensively to assess the probability of corrosion of reinforcing steel in accordance with ASTM C876 [21] as well as for pre-design CP surveys.

In the standard, it is indicated that if the measured potential is less negative than −200 mV versus copper/copper sulphate reference electrode (Cu/CuSO$_4$) there is a greater than 90% probability that no corrosion is occurring. For readings more negative than −350 mV with respect to Cu/CuSO$_4$ there is a greater than 90% probability that corrosion is occurring. For intermediate potential values, the corrosion activity of the reinforcement is uncertain. These values were originally obtained from empirical data obtained from a number of atmospherically RC bridge decks and have provided adequate guidelines for atmospherically RC structures. However, care must be exercised to apply the preceding guideline for assessing corrosion activity for buried substructure owing to the likelihood of reduced oxygen availability, dissolved oxygen being generally the predominant cathodic reaction that fuels the corrosion process. When the access of oxygen to the reinforcement is limited, it may cause the electrochemical potential to shift in the negative direction and may be not represent active corrosion. Such situations may arise for submerged or buried RC structures. The first scenario would suggest that the concrete pores are filled with water and thus restrict oxygen diffusion from the external environment through the cover. The restriction of oxygen at the steel/concrete interface with the second scenario may be caused by exposure to groundwater or water-logged soil, which would render the concrete water saturated, or due to low permeable soil, which would create a barrier between the concrete and the atmosphere restricting oxygen diffusion through the soil. The influence of low oxygen availability is acknowledged by the American Concrete Pressure Pipe Association [15], which states that the corrosion susceptibility for CCP continuously immersed in high chloride electrolyte such as sea water would be very low due to the extremely low rate of oxygen diffusion through the saturated mortar coating.

Very negative potentials, greater than −350 mV quoted in ASTM C876 but with no significant corrosion taking place, are commonly reported [22] ([23] referring to [24],[25] referring to [26]). Elsewhere, however, significant corrosion under oxygen-depleted condition has been reported [37].

A good illustration of the interrelationship between the oxygen availability at the reinforcement and the electrochemical potential is that given by Heuzé [27] who described the natural potential of steel in concrete in dry or atmospherically exposed condition to be in the region of 0–100 mV versus Ag/AgCl compared to −400 to −500 mV in sea water immersion conditions for a concrete offshore platform. As far as buried concrete structures are concerned, the same phenomena is illustrated by the work of Hall [28] where the results of a potential survey conducted on a PCCP gave readings of −600 to −700 mV with respect to Cu/CuSO$_4$. The pipeline was located within the water table with a resistivity of 200–400 Ω-cm. Following excavation at four locations, relatively small corrosion defects were found at the joints, and potential measurement with the reference electrode placed onto the mortar gave a value of -400 mV with respect to $Cu/CuSO_4$, which subsequently depolarised.

In a cathodic prevention application for a RC sea water reservoir structure, Chaudhary [29] reported embedded reference electrode natural readings of -556 to -734 mV and -554 to -948 mV (with respect to Ag/AgCl/ KCl) for buried and sea water submerged elements respectively.

The American Concrete Pressure Pipe Association [15] recommends the implementation of CP for recorded potentials more negative than -350 mV with respect to $Cu/CuSO_4$. However, the decision to apply CP is not solely based on a 'one-off' measurement but is generally preceded by extensive analysis such as soil resistivity and chloride ion content and may be based on data from several potential surveys conducted over a period of time before the pipeline installation. It should also be emphasised that bonding of pipe joints that permits the pipe to soil potential survey is only carried out when there is well-founded suspicion of corrosion activity.

It is further indicated in the document that pipeline potentials more positive than -200 mV with respect to $Cu/CuSO_4$ indicate the pipeline is free of damaging corrosion, in the absence of CP or stray current corrosion.

6.2.2.2.2.4 ROLE OF THE OXYGEN AVAILABILITY

Based on practical corrosion experience for both atmospherically exposed concrete structures and buried CCP, it may be stated with some degree of confidence that chloride-induced corrosion is the most probable cause of attack on buried RC structures. However, it should be emphasised that the rate of attack is governed by the availability and the replenishment of oxygen at the steel/concrete interface. Whereas, in most instances, RC exposed to the atmosphere complies with this requirement, buried concrete by definition may not have the same availability of oxygen. A buried RC structure in an inherently less aerobic environment, exposed to an equivalent level of chloride ion as an above-ground concrete structure, would be characterised by a markedly reduced corrosion rate as illustrated by the schematic Evans diagram in Figure 6.3. Consequently, the time to concrete distress by corrosion is anticipated to be significantly delayed for a buried RC structure when compared to an atmospherically exposed concrete with similar chloride content.

Based on the preceding mechanism, the level of oxygen availability must be strongly borne in mind when using the ASTM C876 standard for the assessment of corrosion probability for buried RC structures. Figure 6.3 indicates a negative shift in the natural corrosion potential (E to E') with a reduction in the corrosion rate (i to i') when the oxygen access is restricted.

The type of soil and its associated electrical resistivity may also affect the dependence of oxygen content for chloride-induced corrosion, further

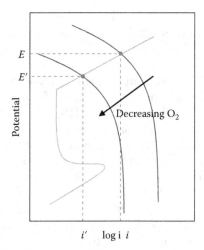

Figure 6.3 Chloride-induced corrosion with various oxygen contents schematic Evans diagram.

complicating interpretation of the results and corrosion appraisal. As previously stated, it is standard practice to consider soil resistivity in the evaluation of the corrosion likelihood of soil and backfill. As Benedict [6] pointed out, highly permeable soil combined with soluble chloride ions at the concrete/soil interface may cause significant corrosion. Among the worst cases is aerated sand that, if not compacted, can exhibit even less favourable resistivities. In such cases, the predominant required cathodic reactant of the corrosion process, oxygen, is in sufficient concentration to cause a significant rate of attack as illustrated in Figure 6.3. Although it is acknowledged that dry sand containing chloride ions is not anticipated to initiate corrosion, particular consideration should be given to situations where the chloride ions are made soluble either by capillarity action from underlying groundwater, the presence of moisture due to high humidity of the climate, or, in the worst case, by intermittent exposure of the structure to groundwater. The latter scenario was appraised by Gummow [30] to be responsible for numerous failures of pre-stressed CCP due to the detrimental wet/dry cycles and associated enhanced chloride diffusion through the mortar cover. It is also worth noting that one notable North African pre-stressed CCP transporting water and laid in a low-lying sand dune failed catastrophically in the 1980s. There have been anecdotal reports of newly grown vegetation along the pipeline route in this arid desert landscape.

Martin [31], in an evaluation of the corrosion and CP of mounded LPG steel tanks, postulated a mechanism that could be applicable in the case of buried concrete structures. Due to the coarse particle size of the sand, backfill drainage of water from rainfall may cause the soluble corrosive salts to leach out in the vicinity of the steel surface. Once the majority of the sand has dried, moist sand may still be in contact with the tank and as air diffuses rapidly through the drained sand any oxygen consumed from the water due to the corrosion process can be rapidly replenished and a rapid rate of pitting may occur.

Experience from the buried steel pipelines industry would suggest that corrosion in aerated sand does not cause significant problems. A possible explanation proposed by Eyre and Lewis [10] is 'The initial corrosion rate in a highly aerated soil is rapid, but the corrosion products can be dense and tightly adherent acting as a protective coating. In a poorly aerated soil the initial corrosion rate is slower and the corrosion reaction does not proceed rapidly enough to form a protective type of corrosion product. Thus in poorly aerated soil, corrosion tends to continue at a relatively constant rate and failure can occur earlier than high aeration. Poor aeration often results in a concentrated pitting type of attack while good aeration usually results in distributed attack with minimal pitting'.

However, the preceding mechanism remains a valid potential cause of corrosion for RC as all the elements favourable to chloride-induced corrosion are supplied. It should be noted that only highly permeable sand (well aerated) could result in a high corrosion rate in a low conductive soil; BS EN 12501-2 [32] provides some guidelines with respect to the type of soil and its degree of permeability (oxygen access) as indicated in Table 6.1.

A convenient technique for assessing the aeration of a particular sand in the field is the use of the oxidation reduction (Redox) potential technique. This technique has been extensively used in the buried steel pipelines industry to assess the probability of microbiologically influenced corrosion caused by the sulphate-reducing bacteria (SRB). The technique permits to evaluate the relative concentration of the reduced versus the oxidised forms of chemical species present in the soil. In an aerobic environment, the oxidised and reduced forms are O_2 and H_2O respectively (due to the photosynthesis action from the flora: $CH_2O + O_2 = CO_2 + H_2O$ [33]), and when O_2 availability declines, the Redox potential (Eh) decreases (becomes less positive). For soils containing SRB, the oxidised and reduced forms are SO_4^{2-} and H_2S respectively. The risk of corrosion due to SRB action is considered severe for Redox potentials less than 100 mV (predominant

Table 6.1 Type of soil

Type of soil	Resistivity range (Ω-cm)	Aeration
Marine mud	300–800	Very low
Clays and silts	500–2,000	Low to very low
Dry non-marine sands	20,000–200,000	High

Source: Extracted from BS EN 12501-2:2003, *Protection of Metallic Materials against Corrosion-Corrosion Likelihood in Soil, Part 2: Low Alloyed and Non Alloyed Ferrous Materials*, Comité Européen de Normalisation (CEN), Brussels, Belgium (2003).

oxidation reduction reaction) but is considered slight for values between 200 and 400 mV [34]. For Redox potentials greater than 200–400 mV, it is anticipated that oxidation, due to the availability of oxygen, will be the predominant reaction. The Redox potential is used in other scientific/engineering disciplines to evaluate the soil respiration related to photosynthesis, and Figure 6.4 gives an indication of the relationship between oxygen availability and Redox potential.

It may be noted that if a piece of steel was buried in this soil and its electrochemical potential monitored, a shift in the more positive direction would be expected due to the enhanced oxygen supply.

The corrosion mechanism based on oxygen availability such as could occur in a high resistivity soil with a high oxygen content must be evaluated on a case-by-case basis and could be perceived as a 'direct' corrosion attack as the short-line corrosion currents between anodic and cathodic areas are to some extent reduced due to the low conductivity of the surrounding soil.

This type of corrosion mechanism may be more likely to occur in the Middle East or Arabian Gulf than in temperate climates as the climatic factors such as high humidity and high temperature and the geological patterns such as saline groundwater, Sabkha and sand provide all the necessary conditions for this form of corrosion to develop and proceed at a significant rate.

Soil resistivity, oxygen availability and the resulting native potentials are also thought to have a significant influence on the CP characteristics such as current density demand and protection potential criteria.

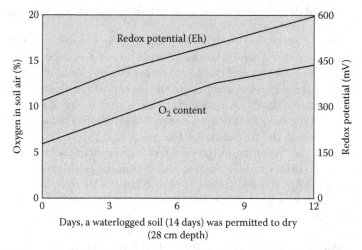

Figure 6.4 Example of relationship between redox potential and oxygen concentration. (www.faculty.plattsburgh.edu/robert.fuller/370%20Files/Weeks13Soil% 20Air%20&%20Temp/Redox.htm, September 2005.)

6.2.2.2.3 Further considerations

Much of the background to the mechanisms of corrosion in buried RC is based on the extensive literature available on the practices of the PCCP industry. However, it should be noted that the intrinsic construction features of RC in foundations and PCCP differ significantly. These disparities, presented in the following, must be taken into consideration when conducting corrosion appraisal on foundations.

PCCP are generally characterised by

- Pre-stressing wire wound around a steel cylinder pipe.
- Mortar coating thickness generally varying from 20 to 40 mm.
- Mortar is factory applied in a controlled environment.
- Quality of the mortar is controlled and water—cement ratio may be as low as 0.3 [35].
- Corrosion failures due to cross-sectional loss of pre-stressing wire are generally noticeable and could result in the bursting of a pipe.

Buried concrete structures and foundation are generally characterised by

- Reinforcement with a greater bar diameter.
- Concrete cover generally in excess of 40 mm for buried applications (and could be as high as 75 mm).
- Due to their loading capacity, water—cement ratios are generally higher than for mortar.
- Concrete mix is generally cast in situ and cured in place and thus the probability of construction problems is higher.
- Structural distress is less noticeable than for above-ground concrete.

Based on the preceding evidence, PCCP usually exhibit a lower cover than for RC structures, which would provide less of a barrier to the diffusion of aggressive species and oxygen than for conventional RC. However, due to their lower water—cement ratio, the ingress of corrosive species and oxygen is further delayed. A reason for this is that lower water—cement ratio decreases the oxygen diffusion due to the associated reduction of concrete pore size and the tortuosity of the pores [22,23]. A drop in oxygen diffusion may also result from the concrete curing where a high degree of hydration may further reduce the size and volume of capillary pores.

However, the pre-stressing wires are more vulnerable to corrosion as less amount of attack is needed for failure. Also, failures of PCCP are more noticeable that it could lead to leaks or, in the worst case, pipe bursts.

Nevertheless, there has been a relatively limited number of failures compared to the total number of PCCP installed [36], and the implementation of CP remains statistically low for such applications and is generally considered as a curative method for defective and existing structures [1,20,36].

Similarly, care must be exercised in the corrosion assessment of existing foundations. However, these structures do not generally contain liquid that could help in the identification of failures, and due to their geometrical configuration, visual inspections for signs of corrosion distress are seldom possible. The use of potential monitoring may provide some indication of corrosion activity for CCP if used to detect any shift from baseline potential. However, a similar technique employed for foundations will not, in most case, lead to conclusive results. These structures are generally more compact and the area of measurement from a reference electrode placed onto the ground surface may not give satisfactory results. Also, one-off potential measurement without the use of a baseline data collected after construction, taken on a structure and indicating very negative readings, may lead to misinterpretation of the actual electrochemical condition of the steel as a result of oxygen availability. It should be stressed that great care must be taken when drawing conclusions from very negative potential readings, may be a delicate exercise leading to a possible 'dual' interpretation as, in most instances, oxygen depletion is synonym to negligible corrosion but it could also be associated with significant corrosion without the production of expansive corrosion products with no associated cracking of the concrete. The following example illustrates the complexity of the corrosion appraisal of buried RC structures:

In March 1998, a form of sulphate attack (thaumasite sulphate attack) was found on the buried supporting RC elements in a 30-year-old motorway bridge in the United Kingdom [37]. The foundations were constructed with reasonable quality concrete, which had been specified in accordance with the current guidance, to cater for the perceived ground sulphate. Evidence was also found of considerable section loss of the reinforcement caused by corrosion with soluble corrosion product leaching out of the concrete on columns and piles buried below the ground.

A particularity of the RC corrosion and CP industry in general is that they are seldom the subject of techno-economic analyses (with the exception of the limited cases of cathodic prevention systems), which could lead to a 'better to prevent than to cure' approach. CP technology has been extensively used in the oil and gas industries to protect bare or coated steel structures in contact with soils and water since the time of construction. It is employed as a methodology of choice for corrosion protection in this industry and is also acknowledged as an economically viable option issued from stringent capital expenditure and operational expenditure analyses that are common in the petroleum sector. Significant maintenance saving is achievable for metallic structures implemented with CP as there is no need for repair or partial replacement of the structures.

Due to the hazardous nature of the fluid transport for most of the oil and gas pipelines, CP in combination with coating is generally implemented to comply with the today's stringent regulations with regard to safety and environment.

It has been reported [1,6] that if the preliminary site investigations reveal that the soil corrosivity is considered to be a problem, relocation of the PCCP route could be considered. Such relocation would be of course rarely possible for the case of foundations or substructure.

Of all the parameters that can be used to assess the corrosion likelihood of a buried RC structure, soil electrical resistivity may be altered before and after construction due to backfill and compaction of the surrounding earth.

Other parameters may be included in the corrosion appraisal of buried RC structures such as low pH; differential aeration concentration cell, which may prevail for piles driven in the earth; electro geological cells between different soils and possibly bacterial influences similar to those encountered in sewer environments, which could result in the degradation of the concrete by the combined action of sulphur-oxidising bacteria causing degradation of the concrete and the subsequent action of SRB on reinforcement exposed to the soil. The latter mechanism is well documented for sewers, but there is little or no documented evidence that this mechanism could prevail for concrete buried in soil.

6.2.3 Selected case histories of CP for non-pipe applications

A selection of several case histories of CP for buried RC structures, excluding the case of CCP, provides an insight into the versatility and, in some cases, complexity of the subject.

6.2.3.1 CP of the Victorian Art Centre piling foundation (Cherry, Melbourne, Australia)

Cherry reported the application of CP from the time of construction for the piling foundations of the Victorian Art Centre in Melbourne, Australia [5]. The RC building consists of 12 floors with 7 of them below ground level. The structure was designed to extend downwards 20 m below the groundwater level. A particular aspect of the structure is that it was designed with bare mild steel tension piles to accommodate the buoyancy forces.

The soil water and groundwater were considered to be very aggressive to both concrete and steel as the sulphate and chloride contents were measured on site to be 1400 mg/L (ppm) and 12,000 mg/L (ppm) respectively.

A temporary zinc galvanic anode CP system was installed on the piles to confer corrosion protection between the time of installation and the energisation of an impressed current CP system. The permanent impressed current CP system consisted of eight deep-well groundbeds rated 80 amps each and distributed across the piling foundation. It is acknowledged that the system capacity could be regarded as excessive for protection of reinforcing steel but it was necessary to also allow for the current demand of the buried bare steel in close vicinity to the structure.

The design criterion for the system was a protection potential of –950 mV with respect to $Cu/CuSO_4$, and after 10 years the potential was of –950 to –1060 mV with respect to $Cu/CuSO_4$ and operating at half of the initial CP system current capacity.

It should be noted that the current demand of the structures was not based on the surface area of the embedded steel and bare steel in soil but on the computed current capacity of the impressed system from the result of the temporary CP system.

6.2.3.2 CP of a reinforced concrete slab foundation (Howell, San Francisco, California)

Corrosion of reinforcing steel in residential slab foundations was identified in several housing developments constructed on chloride-contaminated soil within the San Francisco area [38]. Rust staining and spalling of the concrete developed on the atmospherically exposed areas just above the grade level and in the interior of the basements. Soil analyses revealed average chloride and sulphate ion contents of 2850 and 1230 ppm, respectively, with a soil resistivity of 179 Ω-cm. An impressed current CP system was subsequently installed with distributed anodes. The anodes were distributed close to the foundation to provide current to all areas of the reinforcing steel and with view to limit current pickup at nearby residences due to electrical continuity conferred by the power company earthing.

6.2.3.3 CP of buried pre-stressed concrete tank (Heuzé, Europe)

Heuzé described the application of CP to buried pre-stressed concrete structures and in particular to pre-stressed concrete tanks in the 1960s [12]. Figure 6.5 represents a complete external CP installation carried out at the time of the tank construction with a main underlying anode similar to what

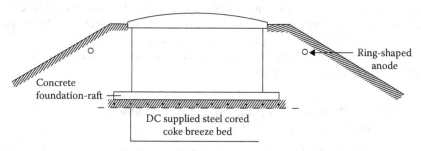

Figure 6.5 Cathodic protection of buried pre-stressed concrete tank. (Heuzé, B., *Corrosion and Cathodic Protection of Steel in Prestressed Concrete Structures*, NACE Western Region Conference, Phoenix, AZ, 1964.)

is now commonly encountered for steel storage tank bottom CP systems. The anode consists of a coke breeze bed providing a uniform field of CP for the entire bottom area due to the high conductivity. An auxiliary peripheral anode is provided for protection of the wall.

6.2.3.4 CP of sewerage pumping station (Das, Middle East)

Das gave account on the corrosion and subsequent CP of two sewerage pumping stations buried in the ground to a depth of 13 m [39]. Reinforcing corrosion was identified within the internal surface of the pump chamber containing raw sewage. The concrete walls were up to 2 m thick with two layers of steel reinforcing bars varying between 16 and 32 mm diameter. The structures were buried in soil with groundwater that was found to have a high chloride content and the external surface was protected with a tanking system. Parts of the internal arrangement 'wet' wells were lined with a proprietary material to protect against raw sewage and the other 'dry' well compartments were not coated and subsequently showed signs of concrete deterioration. However, corrosion of the external concrete face could not be determined without extensive excavation. Conventional CP systems with anodes contained within the structure were not selected due to the need for coating removal and also because it was judged unlikely that such a system could adequately protect the reinforcement at the soil side concrete face.

A CP system using distributed groundbeds around the structure periphery was retained. The system consists of 10 single-anode groundbeds buried at 10 m depth for each pump station substructure with the use of a single transformer rectifier rated at 100 A/15 V. A number of embedded reference electrodes were distributed across the structures (internal and external faces of walls). Almost all recorded native potential ranged from −300 to −500 mV with respect to Ag/AgCl with no significant difference between the internal and external faces. The CP potential criterion adopted in the project was the use of the 100 mV polarisation shift from native potential, and after 6 months of operation all reference electrodes were well in excess of the criterion. Data analysis indicated that the reinforcement closest to the external walls polarised at a very early stage, whereas a considerable time lapse was needed for the reinforcement nearest the internal walls to achieve minimum polarisation.

6.2.3.5 CP of foundation (Chadwick and Chaudhary, Saudi Arabia)

Deterioration of RC foundations was reported in one of the Saudi Arabia's largest refineries [40]. Cracking and deterioration of the concrete structures below ground level up to depth of 3 m were noted after 7 years of operation due to chloride-induced corrosion. Chadwick and Chaudhary gave an account of the results of a CP pilot on a selected foundation structure comprising an octagonal concrete pad measuring 0.8 m depth by 8.93 by 8.93 m side. The structures also comprise concrete columns mostly atmospherically exposed and the top of the slab was 0.3 m below ground level. The top of the slab and the columns were independently cathodically protected using a mixed metal oxide—coated titanium mesh anode and the bottom of the base slab was protected with four silicon iron anodes buried in the soil and distributed around the periphery. It should be emphasised that design calculations undertaken by the investigators indicated that only two anodes were required but four were selected to ensure even current distribution. This approach of CP design simply based on the division of the structure current demand by the output current of a single anode to obtain the number of anodes required will be addressed later. Initial energisation revealed that 35 mA/m^2 of steel was required to obtain a polarisation shift greater than 150 mV and 32 mA/m^2 of steel was also reported to achieve similar polarisation shift for the mesh system of the top of the slab. A polarisation decay of 100 mV decay after 4 hours was observed for the buried slab with the decay gradually decreasing up to 24 hours with the exception of two locations. During the subsequent year of monitoring, carried out at 4-month intervals, the applied current density was gradually decreased to 15 mA/m^2 of reinforcing steel surface area.

6.2.4 CP criteria

By definition, CP criteria are strongly dependent on the environment to which the cathode (the metallic object to receive protection) is exposed. In most instances, the protection potential and current density criteria (current demand to reach the protection potential) rely on the environmental characteristics. All these parameters depend on one another.

The potential and current density criteria for the application of CP for cementitious-coated pipelines and atmospherically exposed concrete have been the subject of intensive research and debate until the application of normative documents in the 1990s for the United States [41] and in 2000 for the European Union [42].

However, while these standards successfully address the requirement for CP applications of above-ground concrete structures, the strict application of their recommended CP criteria for buried applications may not necessarily give adequate results owing to the influence of the soil component.

To comprehensively evaluate the CP criteria for a particular application of buried RC, the experience of both buried steel structures and atmospherically exposed RC with regard to the CP criteria needs to be considered.

A concise write-up of the industry practices with regard to the protection potential for buried steel and above-ground concrete structures are presented in the following sections, supported by relevant case histories of buried concrete CP applications with an emphasis on the resulting current densities.

6.2.4.1 Protection potential

6.2.4.1.1 Buried steel structures

The universally adopted CP potential criterion for buried steel application is −850 mV with respect to $Cu/CuSO_4$ instant OFF in aerobic conditions or −950 mV with respect to $Cu/CuSO_4$ in anaerobic soil where the action of SRB is suspected. It is an empirical potential derived from practical experience of protection of buried steel pipelines and is the value to be reached to shift the structure potential from corrosion to immunity and is thermodynamically supported, in most instances, by the Pourbaix diagram (E-pH) of iron in water. It should be emphasised that the application of the criterion as instant OFF (i.e. neglecting voltage drop on the measurement due to the flow of current through the soil) is now universally utilised, but it was not the case in the past, based on different practices around the world. Wyatt [43] indicated that this matter was well documented and understood in Germany and continental Europe in the late 1960s to early 1970s with separate evidence of an earlier application of this criterion in France dating back to 1964 given by Heuzé [12]. This knowledge was reflected in the United Kingdom and American standards in the mid-1980s and mid-1990s, respectively [43]. This should be borne in mind when reading publications preceding these dates.

There is a secondary criterion to cover particular soil condition of highly resistive and well-aerated sand for buried steel pipelines as indicated in BS EN 12954:2001 [44]. Table 6.2 is an extract of the information given in the standard.

The purpose of the reduced protection potential requirement as a function of the soil resistivity is aimed to solve practical problems encountered in the field with the technical constraints or cost-effectiveness in achieving the −850 mV with respect to $Cu/CuSO_4$ criterion in such soils. Another valid definition of CP is to reduce the external corrosion rate by lowering the steel potential to an economically acceptable level. A conservative natural corrosion potential, valid for the majority of soil encountered along a pipeline route, would be in the order of −600 to −400 mV with respect to $Cu/CuSO_4$ as indicated in Table 6.2. The application of CP would result in a negative shift of 450–250 mV to achieve the instant OFF potential of −850 mV with respect to $Cu/CuSO_4$ and is characterised by a current density reflecting the potential shift and polarisation resistance through Ohm's law. When the instant OFF potential

Table 6.2 Natural and CP protection potentials for various type of soil

Metal or metal alloy		Medium	Free corrosion potential: E_n (without cell formation) indicative value (Volt) (Cu/CuSO$_4$)	Protection potential (Volt) (Cu/CuSO$_4$)
Non-alloy and low alloy Fe materials with yield strength ≤800 N/mm²	Water and soil aerobic conditions	Normal condition ($T \leq 40°C$)	−0.65 to −0.4	−0.85
		Aerated sandy soil ($10,000 < \rho < 100,000$ Ω-cm)	−0.5 to −0.3	−0.75
		Aerated sandy soil ($\rho > 100,000$ Ω-cm)	−0.4 to −0.2	−0.65
	Water and soil anaerobic conditions		−0.8 to −0.65	−0.95

Source: Extracted from BS EN 12954: 2001, *Cathodic Protection of Buried or Immersed Metallic Structures—General Principles and Application for Pipelines*, Comité Européen de Normalisation (CEN), Brussels, Belgium (2001).

of −850 mV with respect to Cu/CuSO$_4$ is applied in highly resistive and well-aerated sand adjacent to conventional soil, this would result in a polarisation shift (based on a native potential varying from −200 to −500 mV with respect to Cu/CuSO$_4$ as per Table 6.2) of 350−650 mV. Based on Ohm's law and neglecting the possible change of the polarisation resistance for this type of soil, it is anticipated that this would result in an increase in the current density demand. This could be only achieved by increasing the CP station outputs that may result in an unnecessary current density demand increase for the majority of the pipeline laid in conventional soil. It may also be perceived that the use of the −850 mV with respect to Cu/CuSO$_4$ criterion is not required in this type of sand and its application may result in an unnecessary current density requirement for adequate protection compared to the use of the secondary protection potential described in Table 6.2.

Complementary to the −850 mV with respect to Cu/CuSO4 potential criterion, National Association of Corrosion Engineers (NACE) also stipulates the use of the 100-mV polarisation shift or decay as a valid approach for CP of buried steel pipelines [45], steel tank bottom or underground storage tanks. The polarisation shift indicates 100-mV difference between the natural corrosion potential and the instant OFF CP potential, while the decay relies on a 100-mV difference between the instant OFF CP potential and the depolarised value after some time. A comprehensive evaluation of the criterion is described elsewhere in the literature [46]. North American practical application of the use of the 100-mV criterion indicated economic advantages for cathodically protecting bare and poorly coated pipelines or pipeline laid in highly resistive soils [47].

A third influencing normative organisation is the International Organisation for Standardisation, which has adopted the European normalisation's (EN) requirements for highly resistive and highly aerated sand in the last revision of their normative document for buried pipelines [48]. The document also acknowledges the use of the 100-mV polarisation shift or decay.

6.2.4.1.2 Atmospherically exposed concrete

BS EN ISO 12696:2000 proposed the following criteria to be achieved to demonstrate successful commissioning and operation of a CP system:

> An instant OFF potential (measured between 0.1 s and 1s after switching the DC circuit open) more negative than –720 mV with respect to Ag/AgCl/0.5 M KCl,
> A potential decay over a maximum period of 24 hours of at least 100 mV from instant OFF,
> A potential decay over an extended period (typically 24 hours or longer) of at least 150 mV from the instant OFF subject to continuing decay.

It also specifies that no instant OFF potential shall be more negative than –1100 mV versus Ag/AgCl/0.5 M KCl reference electrode to prevent detrimental effects to the steel.

For existing RC structures showing distress, the current density required to achieve adequate CP ranges from 2 to 20 mA/m² irrespective of the mechanism responsible for the corrosion. Corrosion caused by carbonation is characterised by uniform corrosion of the reinforcement, whereas chloride-induced corrosion causes localised pitting attacks with the remaining reinforcing steel still in the passive state.

For new RC structure exposed to aggressive environments (cathodic prevention), the document recommends the use of a current density ranging from 0.2 to 2 mA/m².

NACE RP 0290-90 [41] specified the use of a polarisation shift or decay of 100 mV but no specific period after the interruption of the CP current is quoted and is left to the judgement of the CP engineer or practitioner. Similar potential criteria are referred to for the CP of PCCP [49] with a positive limit of –1000 mV with respect to $Cu/CuSO_4$ to prevent hydrogen generation and possible hydrogen embrittlement of the pre-stressing wire. It is also quoted in the latter document that values of polarisation shift or decay less than the 100-mV criterion may be sufficient to provide adequate CP.

6.2.4.2 Current density

Various current densities for CP of buried RC structures are quoted in the literature and the protection potential adopted together with the surrounding environment would typically influence the resulting current density. This is best illustrated by the selection of case histories of CP applications for cementitious-coated pipelines presented in the following discussion.

6.2.4.2.1 CP of PCCP (Unz, Israel, 1960)

An impressed current CP station was implemented on a 1400-m-long section of line fitted with insulating joints at both ends [50]. The line consisted of 36″ pre-stressed concrete pipe of the non-embedded cylinder type, with ¾″ cement mortar coating. All bituminous-coated components including blow-offs and air valves were also fitted with insulating joints. The ground was a saline soil considered corrosive with sulphate content up to 7000 ppm and soil resistivity ranging from 45 to 220 Ω-cm. Some of the sections of the line were in contact with the groundwater and one section was laid above the ground and covered with an embankment.

The impressed current CP system had the particularity to have its drain point located close to one end, and results from the CP system activation in combination with the soil characteristics along the line are described in Figure 6.6.

It can be seen from Figure 6.6 that the natural potentials were in the region of 320–350 mV with respect to Cu/CuSO$_4$, with slightly more negative potential readings in the low resistivity soil around the drain point. The polarisation shift (difference between off and natural potentials) resulting from the application of CP ranged from 280 to 262.5 mV in the vicinity of the drainage point and from 239 to 231 mV near the opposite end of the pipeline.

The resulting average current density was estimated to be 4 mA/m², but figures as high as 11.8 mA/m² in the vicinity of the drainage point and associated groundbed were calculated; however, no information regarding the location of the groundbed with respect to the pipe and its current and voltage output was given in the publication. For pipe section away from the drainage point, current densities (disregarding the value quoted for the above-ground section) ranged from 1.26 mA/m² for pipe sections close to the far end to 1.67 mA/m² for pipe sections located in the middle of the line.

It should be noted that the preceding mentioned current densities were calculated by dividing the current pickup of the relevant section by the respective surface area of the pipe. The pre-stressing wire surface area was not taken into account, which should mean that the 'true' current density based on the total steel surface area would be somewhat lower than the figures expressed.

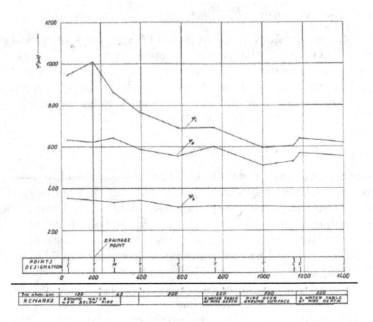

Figure 6.6 On (φ_i), off (φ_p) and natural (φ_b) potential profiles and soil resistivity along the line. (Unz, M., *Corrosion*, 16, 289–297, 1960.)

6.2.4.2.2 CP of 2 no PCCP (Spector, Israel, 1962)

Spector underlined that the use of the widely accepted potential criterion of −850 mV with respect to Cu/CuSO$_4$ for steel main could not be considered as a reasonable criterion for the protection of RC pipe [51]. It was his view that a change in potential of about 300 mV would be a more appropriate criterion in such a case. Due to the possible detrimental effect of overprotecting pre-stressed CCP, Spector was satisfied in the early stage with a 200–300-mV shift as a potential criterion and his view was to change the potential at latter stage once more experience is gained. It would appear that the polarisation shift referred to in the publication is the difference between the natural potential and the ON potential, not the instant OFF potential. This should be taken into account when correlating the current densities and the associated polarisation shifts.

Results from two different applications were presented, reflecting the adopted practices:

6.2.4.2.2.1 TEL-AVIV MUNICIPAL AREA

Impressed current CP system was implemented on a 15-km pre-stressed concrete cylinder line with diameter ranging from 1 to 2.5 m below ground. It would appear that the CP system was installed a relatively short period of time after the pipeline laying. The concrete line was electrically isolated from steel main by the use of isolating joints. No information related to the soil resistivity of the top soil was given but the geological pattern was characterised with the presence of groundwater at approximately 20 m depth. The climate in the region is anticipated to be relatively dry with sand at the pipe depth and sandstone just below.

Natural potentials were recorded and ranged from 0 to −250 mV with respect to Cu/CuSO$_4$. Current drain tests using temporary groundbeds were conducted to determine the current requirement of the concrete pipe. The current densities found necessary for achieving protection by 200–300-mV polarisation shift from the native potential were 3–4 mA/m^2, based on the external surface of the pipe, not the actual steel surface area. It was also reported that 1 mA/m^2 of concrete pipe outside surface area was sufficient to shift the potential by 100–200 mV; this was subsequently used for the installation of magnesium galvanic anodes for some isolated concrete pipe section in proximity to steel main.

Figure 6.7 shows the results of the current drain test, which confirm that the polarisation shift was based on the ON potential rather than the instant OFF. It was also reported that after 3 days of supplying constant current, the ON polarisation potential resulted in an increase of an average of 100 mV, the latter may probably be due to the reduction in protection current demand from initial to stabilised current density requirement.

From Figure 6.7, it can be seen that the natural potential ranged from +50 to 0 mV with respect to Cu/CuSO$_4$. The application of CP resulted in a polarisation shift (between natural and ON potential) of 300 mV at the drainage point location and 100–200 mV elsewhere. From modern CP practices, it could be concluded that less than 3–4 mA/m^2 of steel surface

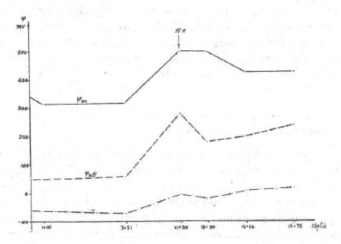

Figure 6.7 Typical current drain test. (Spector, D., *Corr. Technol.*, 9, 257–262, 1962.)

area was necessary to cause a polarisation shift of 100–200 mV for the newly laid CCP with most of the embedded steel undamaged or passive and, based on the natural potential, possibly in an aerated soil strata.

6.2.4.2.2.2 SOUTHERN ISRAEL AGRICULTURAL AREA

Spector provided a second account of an impressed current CP system installation for a pre-stressed CCP. The total length of the pipe was 38 km with diameters ranging from 30″ to 36″ and buried at a depth of between 1 and 2.5 m. The concrete main provided water for irrigation to agricultural settlements in the area. Soil resistivity measurements were conducted along the pipeline route and indicated figures from 650 to over 10,000 Ω-cm and the soil strata was similar to that experienced in Tel-Aviv with a water table located at 30–60 m depth. The concrete main was in operation for about 8 years and had suffered from corrosion in some locations resulting in bursts.

Electrical continuity between the pipe sections was determined and a natural potential survey was conducted with values ranging between about –250 and 500 mV with respect to $Cu/CuSO_4$. It was further observed that the potential difference between corroding and non-corroding locations was in the order of 100–400 mV.

The CP system allows some of the protective current to flow to the well casing and steel mains bonded to the concrete main. The current densities necessary to achieve a 300-mV polarisation shift (it is assumed they are measured between the natural and ON potentials) was on average 20 amps/km for the 30″ main and 60 amps/km or greater for the 36″ main. These are equivalent to 8.3 and 20.1 mA/m^2 (based on the external concrete pipe surface area) for the 30″ and 36″ line respectively. Limited conclusions can be drawn from these results as the CP system was designed to primarily protect the concrete main but also included limited protection to the bare steel surfaces. From the applied current tests, it was also concluded that a groundbed placed at 100 m from the pipe was sufficient to obtain a span of protection of 3 km with 300-mV shift at both ends.

6.2.4.2.3 CP of steel in pre-stressed concrete
(Heuzé, France, 1964–1965)

Heuzé presented a comprehensive evaluation on the corrosion and the CP of steel in buried pre-stressed concrete structures [12,27]. The investigator emphasised that corrosion and failure of such structures (pipelines and tank) are rare and the use of CP should always be considered as a remedial measure rather a preventive one (from the time of construction) based on the inherently good properties of the mortar. Based on the experience from continental Europe, Heuzé relied on cathodically protecting steel in concrete in a similar manner than for steel in soil, that is, shift the steel potential into the immune region by aiming for −900 to −1000 mV with respect to Cu/CuSO$_4$ (instant OFF). It was suggested that the current densities to achieve the preceding protective potential relied more on the concrete quality than the surrounding soil and quoted figures of 0.4–1 mA/m^2 for a good covering concrete such as monolithic or high cycle vibrated or 1–2 mA/m^2 for a middle or mediocre quality, but also reported values in excess of 4 mA/m^2 for defective mortar coating in aerated soils of a mildly aggressive nature. All current densities expressed refer to the steel surface area (i.e. cylinder and wires), not the external surface of the pipe. The author gave details of both impressed current and magnesium galvanic anode CP systems and relied on the use of bonding cables to each of the electrically discontinuous pipe segments rather than bonding all pipe segments together. It was thought that this method would provide better results in the current attenuation and more even distribution of the protection potential. Spacing between the impressed current stations was anticipated to be 4–6 km.

6.2.4.2.4 CP of mortar-coated steel pipeline
(Deskins, California, 1966)

Deskins presented a number of CP applications for mortar-coated steel main in an agricultural region of the southern Californian coast [52]. The soils ranged from low-lying beach sand saturated with sea water to freshwater in irrigated regions. Due to electrical continuity between organically coated and mortar-coated steel pipeline of the water main, Deskins proposed the use of −600 mV with respect to Cu/CuSO$_4$ as a compromise CP potential criterion for economic reason. The investigator acknowledged that this criterion would not confer total protection to the organically coated pipelines but was considered to adequately protect the majority of the water mains that were mortar-coated steel.

Deskins gave details of a CP system installed on 2700 ft. of isolated mortar-coated steel. The steel cylinder was 6⅝″ in diameter and had some organically coated fittings. Soil resistivity varied from 1200 to 20,000 Ω-cm. The potential profile with the CP system interrupted under a switching cycle of 20 seconds on and 40 seconds off were recorded, as shown in Figure 6.8.

The recorded natural potentials ranged from −140 to −190 mV with respect to Cu/CuSO$_4$, and as much as 200 mV polarisation (from natural to instant OFF reading) was achieved approximately at ½ mile from the current source with instant OFF (considered as 'semi' instant OFF due to the switching cycle, true instant OFF being possibly more electronegative), varied from −250 to −400 mV with respect to Cu/CuSO$_4$. The computed current density necessary to achieve the required polarisation, based on the external surface of the pipe (not including the reinforcement wire mesh or other fittings), was 4.1 mA/m^2.

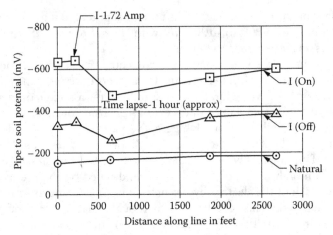

Figure 6.8 Potential profile. (Deskins, R. L., *Mater. Protect.*, 5, 35, 1966.)

Deskins also gave an account of the CP of two other pipes protected with magnesium anodes installed in soil of low resistivity with current densities of 2.33 and 3.3 mA/m². However, limited information regarding the natural potential and OFF potential reading were given, only ON potentials from –700 to –900 mV with respect to Cu/CuSO$_4$ were reported, with potentials as high as –1020 mV with respect to Cu/CuSO$_4$ at the drain point. The current densities were obtained under controlled conditions where the mortar coating was known to be in good condition and no service line and other fixtures connected.

6.2.4.2.5 CP of PCCP (Gourlet, Australia, 1978)

Gourlet claimed to have implemented the first application of CP for prestressed concrete pipeline in Australia by referring to the Ross River pipeline at Townsville in Queensland [19]. The author conducted a survey of the literature available at the time and concluded that PCCP could be satisfactorily cathodically protected by adopting a 300-mV shift from the natural potential when the CP system is initially switched on and relying on a final potential between –450 and –1100 mV with respect to Cu/CuSO$_4$ (it is not clear if the latter potentials are ON or instant OFF, but from the more negative figure expressed it would appear that they are ON values).

The PCCP was 9 km long and 1.22 m in diameter. The soil resistivities measured along the pipeline were considered to represent a very corrosive environment with 13% less than 1000 Ω-cm, 80% less 5000 Ω-cm and 97% less than 10,000 Ω-cm. A CP system utilising zinc galvanic anode was installed just after construction (i.e. corrosion prevented from the time of construction). A natural potential survey was conducted along the line and readings ranged from –100 to –250 mV with respect to $Cu/CuSO_4$. Measurements made immediately after anode connection indicated that the 300-mV shift criterion was achieved and subsequent potential surveys were carried out at 83 and 230 days, with the data summarised in Table 6.3.

By measurement of current flow from anode banks, it was determined that the operating current density was 1 mA/m^2 of embedded steel. However, it should be noted that these measurements are well documented after 83 and 230 days but initial current values are not presented in the document, that is, these figures may refer to a stabilised current density rather than the initial. Also, these values are associated with CP applied to a pipe free from corrosion.

It can be noted from the preceding data that an increase of protection potential between 83 and 230 days survey is not accompanied with a change of current. It should be emphasised that the 230-day survey was conducted when the soil was wet and it could be expected that there was a drop in the resistance to earth of the galvanic anode bank and possibly an associated drop in the pipe leakage resistance.

An interesting recommendation made by Gourlet, based on the experience of the Townsville pipeline, was that CP would be installed on all future pre-stressed pipelines irrespective of soil corrosiveness.

6.2.4.2.6 CP of 10 concrete coated steel pipelines (Deskins, California, 1979)

Deskins presented data from CP applications on 10 concrete-coated steel pipelines across California [53]. The adopted criterion was a fixed potential of –500 mV with respect to $Cu/CuSO_4$ (assumed ON) provided an assurance could be given that the concrete coating was undamaged, or –850 mV with respect to $Cu/CuSO_4$ (assumed ON) if it was suspected that concrete coating defects were present. Pipe to soil potential survey data was obtained on 10 cement-coated steel cylinder pipelines all laid in soil of resistivity less than 3000 Ω-cm. Some were laid in saltwater marshes or are submarine pipelines. All data was obtained when the pipelines were relatively new and natural potentials ranged from –400 to –600 mV with respect to $Cu/CuSO_4$ for half of them and from –100 to –300 mV with respect to $Cu/CuSO_4$ for the remainder, possibly indicating their associated immersed or buried service.

Results from the CP application indicated current density to be 2–5 mA/m^2 with some of the higher current densities obtained on new and apparently undamaged concrete pipes, possibly those that exhibited natural potentials of –100 to –300 mV with respect to $Cu/CuSO_4$ with subsequent CP application at –500 mV with respect to $Cu/CuSO_4$.

Table 6.3 Potential survey: Ross River pipeline (Townsville)

Test point	Soil resistivity (Ω-cm)	Steel to soil potential difference V_{ss} (mV)				Current flow I_z (mA)		Current demand $j(mA/m^2)$	
		Before	Change after 3 days	83 days (Dry)	230 days (Wet)	83 days	230 days	83 days	230 days
23-C	2800	−134	338	−555	−838	458	212	1.7	0.8
35-A	900	−224	323	−640	−802	225	134	0.8	0.5
51-A	600	−238	532	−622	−887	333	165	1.2	0.6
60-B	1800	−144	31	−496	−720	226	239	0.8	0.9
65-C	3000	−130	485	−510	−597	239	226	0.9	0.8
78-	4000	−155	273	−479	−605	No anodes			
89-B	1200	−114	414	−632	−715	448	394	1.7	1.5
Mean	–	−163	385	−562	−738	322	228	1.2	0.9
Design Specification	–	−200	◄−300	◄−450	◄−450	270	270	1.0	1.0

Source: Gourlet, J.T., Corr Aust, 3, 1, 1978.

6.2.4.2.7 *General experience with corrosion and CP of PCCP (Benedict, 1990)*

Various guidelines and technical approaches were expressed by Benedict with regard to soil corrosivity and subsequent implementation of CP for PCCP were presented [6]. Benedict has also given a review of the potential criteria used for CCP and acknowledged that 100-mV polarisation shift (from baseline potential to instant OFF potential) was used with no reported problems for more than 10 years. He also reported other protection potential criteria, among them −850 mV with respect to Cu/CuSO$_4$, which was considered to be adequate only for severely damaged mortar coatings and large areas of depassivated reinforcement in contact with the surrounding soil. Typical current densities reported by various investigators range from 0.02 to 2.5 mA/m^2, supposedly resulting from the application of the 100-mV shift criteria. This data was assumed to be based on stabilised rather than initial current densities.

An account is given of an interesting case history where a concrete cylinder pipeline was cathodically protected with the −850-mV criterion, which resulted in a current density of less than 0.6 mA/m^2 for an immersed line segment and 10–50 mA/m^2 for another line segment buried in a slightly moist, well-aerated sandy soil. Benedict explained this increase of current demand in terms of the influence of oxygen availability.

6.2.4.2.8 *NF A 05-611* CP of reinforced concrete structures: buried and immersed structures *(French Standard, 1992)*

It should be noted that this French standard was withdrawn a number of years ago as a result of the harmonisation of the various European CP practices with view to produce a unified European standard [54]. It is also stated in the document that at the time of its production and implementation, there was no international standards addressing the subject currently in existence.

The document quoted that to cathodically protect a buried RC structure to an instant OFF protection potential of −850 mV with respect to Cu/CuSO$_4$, current densities from 0.5 to 5 mA/m^2 of steel reinforcement surface area could be envisaged but a current density as high as 10 mA/m^2 could also prevail if the concrete was extensively deteriorated or if the external electrolyte was well aerated.

It was also stated that the extent of the CP action is not limited to the external reinforcement steel mat but also affected deeper reinforcement as the throw of CP current could be of a depth equal or more than the first steel layer spacing. This could be verified in the first instance by the use of models.

6.2.4.2.9 CP of existing PCCP (Peris and Guillen, Madrid, Spain, 1995)

Corroded wires caused the failure in two sections of the PCCP making up part of the Madrid water supply network in what could be considered as an arid environment [55]. About 15% of the total pipeline was tested and 40% of the surveyed line indicated general corrosion. A CP system was installed using magnesium anodes, probably to overcome the high resistivity, on 800 pipe sections. Most of the pipe segments were fitted with a single magnesium anode.

The target protection potential criterion was –850 mV with respect to $Cu/CuSO_4$, which required 1 mA/m^2. The natural and protection potentials surveyed after 2 months are presented in Figure 6.9. It would appear that only the pre-stressing wire received protection and the inner embedded steel cylinder was not included in the CP scheme.

The natural potential of any particular section ranged from –100 to around –400 mV with respect to $Cu/CuSO_4$ and correlated protection potentials varied from –750 to –1000 mV with respect to $Cu/CuSO_4$.

Completely stabilised protection potentials surveyed after 6 months ranged from –850 to –1080 mV with respect to $Cu/CuSO_4$ (assumed instant OFF) and no figures related to the current density were given.

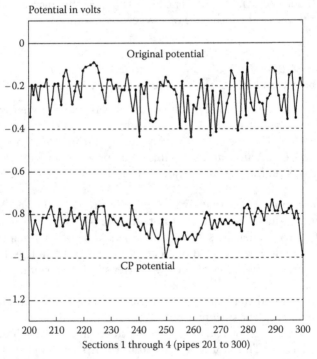

Figure 6.9 Natural and protection potential profile after 2 months. (Peris, M. G., and Guillen, M. A., *Mater. Perform.*, 34, 25, 1995.)

6.2.4.2.10 CP of 2 PCCP (Benedict et al., California, 1997)

The investigators reported the corrosion failure and subsequent CP instal-lation utilising zinc galvanic anodes on two pre-stressed concrete cylin-der pipelines, the Cedar Creek and the Richland Chambers water lines, in California, United States [56].

As far as the design procedure was concerned, the galvanic anode sys-tems were designed to shift the potential to an instant OFF value of between −720 and −820 mV with respect to $Cu/CuSO_4$ to safeguard from hydrogen embrittlement failure of the pre-stressing wire, but a project specification requirement to ensure adequate monitoring performance of the CP system was a minimum 100-mV negative potential shift based on the instant OFF protection potential.

6.2.4.2.10.1 CEDAR CREEK PIPELINE

Most areas of the pipeline contained heavy plastic clays with soil resistivi-ties around 200 Ω-cm. At stream and river crossings, the pipeline was bur-ied in alluvial sands and gravels with a resistivity of up to 12,000 Ω-cm. A trial CP system utilising zinc galvanic anodes was installed on a 4-km section of the line where as many as seven failures had occurred. The natu-ral potential survey gave readings in the region of −450 to −500 mV with respect to $Cu/CuSO_4$ and instant OFF protection potentials in the range −850 mV to −900 mV with respect to $Cu/CuSO_4$, as shown in Figure 6.10.

A minimum polarisation shift of 300 mV was observed on the entire CP system of the Cedar Creek pipeline. Initial current densities ranged from 1.1 to 2.2 mA/m² and average current densities between 0.09 and 0.55 mA/m².

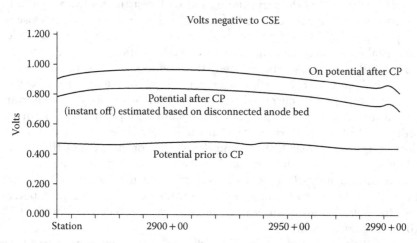

Figure 6.10 Cedar Creek potential survey before and after CP. (Benedict, R. L., et al., *Mater. Perform.*, 36, 12, 1997.)

Other sections of the Cedar Creek pipeline were equipped with similar CP systems and current densities from 0.22 to 1.9 mA/m² (quoted as design, therefore initial) were reported.

6.2.4.2.10.2 RICHLAND CHAMBERS PIPELINE

A zinc galvanic anode CP system was installed on the line based on the same design and operation philosophy. The galvanic anode groundbeds were installed in soil resistivities of 200–1000 Ω-cm. Recorded initial design current densities ranged from 0.83 to 1.66 mA/m².

6.2.4.2.11 Results from a CP pilot on PCCP

Hall (1998) presented an extensive evaluation on the subject of CP of prestressed concrete cylinder pipeline with emphasis on the negative limit applied CP potential to avoid hydrogen embrittlement of the pre-stressing wire, current densities to achieved the 100-mV shift polarisation criterion supported by experiment and review of a number of case histories across the United States [36].

6.2.4.2.11.1 CP PILOT

A number of uncorroded PCCP sections were buried in an arid environment in California with resistivities of 100,000–200,000 Ω-cm dry and of 16,000–30,500 Ω-cm wet. Measurement of the pH and water-soluble chloride and sulphate ion contents revealed that the soil was not considered as corrosive. The pipeline was backfilled with local sand of similar chemical characteristics as the surrounding soil with resistivities from 13,400 to 63,200 Ω-cm. The CP consisted of steel pipe laid perpendicularly close to the pipe at one end with a power supply to simulate situations where anodes are installed close to the pipe due to space limitation. Baseline potential was recorded and was approximately 0 mV with respect to $Cu/CuSO_4$, as shown in Figure 6.11. Initially, the current output of the power supply was set to correspond to a current density of 0.13 mA/m² based on the mortar-coated surface area. After 3 weeks, the instant OFF potentials were in the order of –120 mV with respect to $Cu/CuSO_4$, equivalent to a polarisation shift of approximately 120 mV and depolarisation decay was 100 mV in 4 hours and 120 mV after 1 week. At 6 months, the instant OFF potentials were approximately –220 mV with respect to $Cu/CuSO_4$ and –130 mV after 14 months. The corresponding IR drop from ON to instant OFF potentials for 3 weeks, 6 months and 14 months were, respectively, 50–70 mV, 10–50 mV and 300 mV.

The CP system was subsequently adjusted to simulate an instant OFF potential of –1000 mV with respect to $Cu/CuSO_4$ by applying 1.08 mA/m².

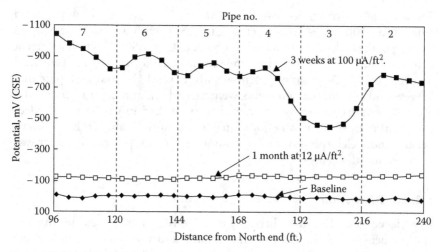

Figure 6.11 Natural and polarisation potentials. (Hall, S. C., *Cathodic Protection Criteria for Pre-stressed Concrete Pipe—An Update*, NACE International, Houston, TX, 1998.)

Some of the potential profiles from the tests conducted are depicted in Figure 6.11.

Other tests were conducted on pipe sections coated with coal tar epoxy and polyethylene encased with pinholes present on both coatings.

CP applied on the coal tar epoxy–coated pipe section indicated that 0.032 mA/m² was sufficient to produce a polarisation shift of 150 mV after 2 months, polarisation decays carried out at 4 hours and 1 month were 65 and 75 mV respectively. Hall suggested that the depolarisation of a coated pipeline may be expected to take longer than for an uncoated pipeline since the diffusion rate of oxygen is low.

A current density of 0.032 mA/m² produces a polarisation shift of 100 mV at 3 hours for the polyethylene-encased CCP and remained essentially unchanged during the month of CP. It produced a depolarisation shift of 140 mV in 4 hours and 200 mV in 1 week.

Tests were also conducted to determine the proportion of the current flowing onto the pre-stressing wire and the cylinder. It was found that 47–49% of the current flow was onto the pre-stressing wire and the remaining onto the underlying steel cylinder. The pre-stressing wire surface area was 65% of the cylinder surface area.

6.2.4.2.12 CP of PCCP in desert area (Garman and Al-Maadani, Libya, 2005)

Garman and Al-Maadani gave details of CP applied to a newly laid 4-m-diameter PCCP in the North African desert using zinc galvanic anodes [57]. Extensive measurements were carried out on a 5.1-km-long pipeline section.

The average soil resistivity was 18,000 Ω-cm. The natural potentials varied from 0 to –100 mV with respect to $Cu/CuSO_4$ with an apparent average reading in the order of –50 mV with respect to $Cu/CuSO_4$. An initial current density of 0.65 mA/m^2 based on the mortar-coated surface area produced an average 271-mV negative potential shift. Several ON and OFF potential surveys and current measurements were carried out at regular intervals and after 18 months an average current density of 0.15 mA/m^2 was recorded. During this interval, it was noted that both the ON and OFF potentials became more electropositive but a minimum 150-mV polarisation shift was recorded for all surveys.

6.2.4.3 Discussion

Experience from the atmospherically exposed RC CP field indicates that the current density for the CP of contaminated and non-contaminated structures exposed to a chloride-rich environment significantly differs based on polarisation decay criteria. A figure of 2–20 mA/m^2 is normally required for the first case and 0.2–2 mA/m^2 (sometimes a figure of 5 mA/m^2 is used for Middle East cathodic prevention applications) for the second one that takes account of the influence of the amount of aggressive species at the reinforcement depth on the corrosion and CP application in generally aerated medium. It should also be noted that these applications rely on impressed current CP system with anodes placed within the structures. The proximity of the anodes combined with a level of care to limit the anodic current discharge in the surrounding concrete may have some influence on the quoted figures. The 100-mV polarisation decay is widely used for such applications and had been proven satisfactory, notably for chloride-induced corrosion characterised by localised pits with a large surface area of reinforcement remaining passive.

For CP applications, it was reported that after initial polarisation of structures at 20 mA/m^2, the system is normally expected to operate at 10%–30% of the rated capacity [5]. However, a reduction of 50% would represent a more conservative value for most applications.

Based on the documented evidence of CP for buried RC structures, attempt to answer the following questions is included in the following text:

- What is the most suitable protection potential criterion (and its influence on the current demand requirement)?
- Is there strong evidence of a reduction of current demand from initial to maintenance period?
- Is there a difference in current density between contaminated concrete with corroding reinforcement and non-contaminated concrete?
- Is there a noticeable influence of the surrounding soil (aeration, resistivity, wet versus dry and natural potential, which are all linked together to a certain extent) on the CP criteria?

From the case histories presented, it can be observed that the majority of the applications rely on the use of galvanic anode system rather than impressed current. It should be emphasised that with an impressed current system, the CP is fully controlled and both the potential and current can be monitored and adjusted. However, the overall operating CP current density is anticipated to be higher compare to galvanic anode system due to increased current density on the line facing the groundbed. Conversely, galvanic anode systems rely on the driving voltage between the anode closed circuit potential and the protection potential. The current output is strongly a function of all the resistance components in the CP circuit and is affected by the surrounding environment. This should be borne in mind in Section 6.2.4.3.1.

6.2.4.3.1 Protection potential

It can be stated that all the presented CP applications exhibited a 100-mV polarisation shift from the natural potential and would consequently comply with one of the NACE requirements; however, from an European CP practitioner point of view, this does not necessarily give evidence of an adequately operating CP system. The application of a 'fixed' potential criterion such as -850 mV with respect to $Cu/CuSO_4$ (Heuzé and NF A 05-611), the ones given by Deskins (-600 and 500 mV with respect to $Cu/CuSO_4$) and, to a certain extent, the use of 300 mV (Unz and Spector) achieved by impressed current or magnesium galvanic anode CP systems resulted in the higher current densities quoted. It should be emphasised that the determination of current density from the application of impressed current systems may result in figures in excess than what is actually required owing to the current spread attenuation and the highest current densities obtained on the line facing the groundbed (even for remote groundbed application). A good example of this was given by Unz.

The application of the -850 mV criterion may result in a potential shift from passivity to immunity, which is achieved by the dissolution of the superficial oxide film. It has also been quoted by several investigators (Heuzé, NF A 05-611 and Benedict) that the application of -850 mV with respect to $Cu/CuSO_4$ in aerated environment results in a high current density requirement. The explanation could be twofold:

- It has been shown previously that in well-aerated concrete the corrosion rate may be several orders of magnitude higher than for the situation with less oxygen. The principle of CP is that sufficient current in excess of the corrosion current is applied to stop the process.
- A second argument would be that well-oxygenated concretes are expected to exhibit more positive natural potential and thus would result in a higher initial polarisation shift from the natural to -850 mV

with respect to Cu/CuSO$_4$ than for low-oxygenated concrete (natural potential more negative). At equivalent circuit resistance between the two conditions and applying Ohm's law, this may result in higher current demand.

Deskins also quoted higher figures on new and apparently undamaged concrete pipes applying −500 mV with respect to Cu/CuSO$_4$. This could possibly be explained similarly as it could be expected that undamaged concrete would exhibit a more positive natural potential than for corroded or damaged concrete.

It could be concluded that, as with above-ground concrete, the use of −850 mV with respect to Cu/CuSO$_4$ for buried concrete CP application is unnecessary and uneconomic solely for the protection of the embedded reinforcement. However, situations may arise where bare steel is in contact with the RC, and in this case, the use of the potential criterion would be valid and consequently result in higher current density requirement not only due to the bare steel but also as a result of such a negative polarisation level for the reinforcement.

In the literature, there is little account on the use of the 100-mV polarisation decay criterion; however, Hall presented data related to a coal tar epoxy–coated pipe, which initially demonstrated compliance with the 100-mV polarisation shift but decayed less than 100 mV after 1 month of depolarisation. Hall attributed this to the lack of oxygen. On the contrary, uncoated CCP buried in the same soil gave satisfactory results. It should be noted that the soil backfill used was sand and therefore it could provide a sufficient oxygen concentration gradient due to its permeability (if not compacted).

The use of 100-mV depolarisation decay in 4 or 24 hours or 150 mV in more than 24-hour decay has been adequate for atmospherically exposed concrete where the oxygen availability is clearly high. However, there is limited documentation on the use of such criteria for buried concrete structures in less permeable soil with high moisture content where the oxygen gradient is anticipated to be much lower.

6.2.4.3.2 Reduction in current demand with time

Benedict, Benedict et al., and Garman and Al-Maadani gave accounts of the reduction of the current density demand from initial to maintenance polarisation [6,56,57]. However, the type of system employed in each case was a galvanic anode system and not impressed current. Careful consideration should be given to the conclusions drawn from the results of the application of a galvanic anode CP system. This is due to the fact that, by definition, galvanic anodes CP systems rely on the electromotive force (driving voltage) between a fixed anode close circuit potential (or operating potential) and the steel potential. Subsequently, at the closure of the CP circuit (anode connected to the protection object), the driving voltage is high (natural potential), and once the steel is polarised to a more negative value, it has diminished until a steady state level dependent on all resistance components prevailing in the CP circuit has been achieved.

Measuring current flowing during the preceding dynamic polarisation process may lead to conclusion where the initial current density is greater by several orders of magnitude than the stabilised current density. Measurement of the current once the steady state potential is reached may be more appropriate and a reduction of the value similarly to what is normally encountered with aboveground concrete CP applications would be anticipated due to the re-alkalisation and chloride removal at the corroded sites, possibly resulting in an increase in the steel surface polarisation resistance.

6.2.4.3.3 Corroded versus non-corroded

One might expect, based on the experience from atmospherically exposed RC, to observe a difference in the current density demand. However, this characteristic mainly results from the application of an impressed current system and based on the application of the 100-mV polarisation decay (i.e. the protection potential is determined and adjusted based on the results of the decay, not based on the result of instant OFF reading alone). As most of the PCCP CP applications rely on the use of galvanic anodes, and therefore the potential cannot be adjusted, it could be expected that most of the systems exhibit protection potentials in excess of what it would be required from a 100-mV decay test. The complexity in drawing firm conclusions with regard to the existence of a change of current density from non-contaminated or corroded buried concrete could be illustrated as follows.

Based on the experimental data presented by Hall, the documented CP trial may be considered as an application of CP on an uncorroded buried concrete (no evidence that corrosion was occurring, value of natural potential in an assumed fairly well-aerated sand). The amount of polarisation was controlled by means of an impressed current system and 0.13 mA/m^2 was needed to maintain a polarisation shift from the natural potential of 120 mV and supported by a polarisation decay of 100 mV.

Gourlet indicated an application of CP for non-corroding concrete pipe and 1 mA/m^2 was required to cause a shift of 300 mV from a natural potential of –100 to –250 mV with respect to $Cu/CuSO_4$ in a low resistivity soil. An interesting feature of this case history is that the current demand remained unchanged from a dry to wet environment whereas in the wet condition the polarisation potential was more negative.

From the Garman and Al-Maadani case study, it could be assumed that the concrete pipe was possibly not affected by extensive corrosion in the supposedly well-aerated desert sand (with a natural potential of –50 mV with respect to $Cu/CuSO_4$). An initial polarisation shift of 271 mV associated with 0.65 mA/m^2 was recorded. After 18 months, the pipe potential was maintained greater than 150 mV from the natural potential with 0.15 mA/m^2; however, the various potential surveys revealed that both ON and OFF potentials became more positive. One could argue that the resistance to earth of the anode groundbed is increasing due to the possible drying of the surrounding soil/anode backfill.

The first example given by Spector could be perceived as an application of CP before initiation of corrosion. The system employed was an impressed current system. Current density in the order of 3–4 mA/m^2 (pipe external surface) was required for polarisation shift of 200–300 mV but based on the ON potential reading and 1 mA/m^2 caused a shift of 100–200 mV. As indicated earlier, there may be evidence of a reduction in current density with time. The top soil was presumably a mixture of sand and sandstone, and natural potential ranged from 0 to –250 mV with respect to $Cu/CuSO_4$.

6.2.5 CP design procedures and prerequisites

6.2.5.1 General

6.2.5.2 CP design prerequisite computations

1. Steel surface area and current demand
2. Anode/groundbed resistance to remote earth
3. Structure resistance to remote earth (leakage resistance)
4. Influence of the anode-structure separation
5. Miscellaneous resistances

REFERENCES

1. Franquin, J., *Protection Cathodique des Tuyaux en Béton Précontraint* [*Cathodic Protection of Prestressed Concrete Pipes*], International Standing Committee for the Study of Corrosion and Protection of Underground Pipelines, Subject 1(b), New York (1972).
2. Barry, D. L., *Material Durability in Aggressive Ground*, Construction Industry Research and Information Association (Ciria) Report 98, Ciria (1993).
3. Concrete *in Sulphate-Bearing Soils and Groundwater*, Building Research Establishment Digest 250 (1981).
4. Morgan, T. D. B., Some Comments on Reinforcement in Stagnating Saline Environments, In *Corrosion of Reinforcement in Concrete*, Ed. Page C. L., Treadaway K. W. J., Bamford P. B., Elsevier Applied Science, London, UK (1990): pp. 29–38.
5. Cherry, B. W., *Cathodic Protection of Underground Reinforced Concrete Structures*, In *Cathodic Protection Theory and Practice, 2nd International Conference*, Ellis Horwood, Stratford upon Avon (June 1989): pp. 326–350.

6. Benedict, R. L., Corrosion Protection of Concrete Cylinder Pipe, *Materials Performance*, 29, 2 (1990): p. 22.
7. Walker, M., *Guide to the Construction of Reinforced Concrete in the Arabian Peninsula*, Ciria Report Publication C577, Concrete Society Special Publication CS 136, Crowthorne, Berkshire, UK (2002).
8. Nicholson, P., *Corrosion and Cathodic Protection of Reinforced Concrete Structures in the Middle East*, Corrosion 96, Paper No 301, NACE International, Houston, TX (1996).
9. Swamy, R. N., *Design—The Key to Concrete Material Durability and Structural Integrity*, International Conference on Reinforced Concrete Materials in Hot Climates, Volume 1, American Concrete Institute, Al-Ain (1994): pp. 3–36.
10. Eyre, D., Lewis D. A., *Soil Corrosivity Assessment*, Transport and Road Research Laboratory, Contractor Report 54, Crowthorne, Berkshire, UK (1987).
11. MOD Technical Bulletin 99-23, Ministry of Defence (MOD), United Kingdom (1999).
12. Heuzé, B., *Corrosion and Cathodic Protection of Steel in Prestressed Concrete Structures*, NACE Western Region Conference, Phoenix, AZ, October 27–29, NACE (1964).
13. *La Protection Cathodique: Guide Pratique*, Chambre Syndicale de la Recherche et de la Production du Pétrole et du Gaz Naturel, Edition Technip, Paris (1986).
14. Gourlet, J. T., Moresco, F. E., *The Sacrificial Cathodic Protection of Pre-stressed Concrete*, Corrosion 87, Paper No 318, NACE International, Houston, TX (1987).
15. *External Protection of Concrete Cylinder Pipe*, American Concrete Pressure Pipe Association, Vienna, VA.
16. Feliu, S., Gonzáles, J. A., Andrade, A., Effect of Current Distribution on Corrosion Rate Measurements in Reinforced Concrete, *Corrosion*, 51,1 (1995): pp. 79–86.
17. BS EN 7430:1998, *Code of Practice for Earthing*, British Standard Institution (BSI), London, UK (1998).
18. Fagan, E. J., Lee, R. H., The Use of Concrete Enclosed Reinforcing Rods as Grounding Electrodes, *IEEE Transactions on Industry and General Applications*, IGA-6, 4 (July/August 1970): p. 337.
19. Gourlet, J. T., The Cathodic Protection of Prestressed Concrete Pipelines, *Corrosion Australasia*, 3, 1 (1978): p. 4.
20. Heuzé, B., Cathodic Protection of Steel in Prestressed Concrete, *Materials Protection*, 4, 11 (1965): p. 57.
21. ASTM C876, *Standard Test Method for Half-Cell Potentials of Uncoated Reinforcing Steel in Concrete*, American Society for Testing and Materials (ASTM), Philadelphia, PA (1986).
22. Design and Operational Guidance on Cathodic Protection of Offshore Structures, Subsea Installations and Pipelines, The Marine Technology Directorate Limited, *Section 7 Cathodic Protection Systems for Concrete Offshore Structures*, MTD Limited Publication 90/102 (1990).
23. Funahashi, M., Bushman, J. B., Technical Review of 100 mV Polarization Shift Criterion for Reinforcing Steel in Concrete, *Corrosion*, 47, 5 (1991): pp. 376–386.

24. Arup, H., Steel in concrete, *Electrochemistry and Corrosion Newsletter*, 2 (1979): p. 8.
25. John, G., *Corrosion Inspection and Cathodic Protection of Reinforced Concrete Structures*, CAPCIS Ltd., Manchester, UK (March 1996).
26. Arup, A., *Cathodic Protection of Steel in Concrete, Korrosioncentralen Newsletter*, 1 (1978).
27. Heuzé, B., Cathodic Protection on Concrete Offshore Platforms, *Materials Performance*, 19, 5 (1980): p. 24.
28. Hall, S. C., Mathew, I., Sheng, Q., *Prestressed Concrete Pipe Corrosion Research: A Summary of a Decade of Activity*, Corrosion 96, Paper No 330, NACE International, Houston, TX (1996).
29. Chaudhary, Z., *Cathodic Protection of New Seawater Concrete Structures in Petrochemical Plants*, Corrosion 2002, Paper No 260, NACE International, Houston, TX (2002).
30. Gummow, R. A., Corrosion and Cathodic Protection of Prestressed Concrete Cylinder Pipe, *Materials Performance*, 44, 5 (2005): p. 32.
31. Martin, B. A., Corrosion and Protection of Mounded LPG Tanks, *Materials Performance*, 29, 9 (1990): p. 13.
32. BS EN 12501-2:2003, *Protection of Metallic Materials against Corrosion-Corrosion Likelihood in Soil, Part 2: Low Alloyed and Non Alloyed Ferrous Materials*, Comité Européen de Normalisation (CEN), Brussels, Belgium (2003).
33. www.faculty.plattsburgh.edu/robert.fuller/370%20Files/Weeks13Soil%20Air%20&%20Temp/Redox.htm (September 2005).
34. BS EN 7361-1:1991, *Cathodic Protection, Part 1: Code of Practice for Land and Marine Application*, British Standard Institution (BSI), London, UK (1991).
35. Bianchetti, R. L., Corrosion and Corrosion Control of Prestressed Concrete Cylinder Pipeline—A Review, *Materials Performance*, 32, 8 (1993): p. 62.
36. Hall, S. C., *Cathodic Protection Criteria for Prestressed Concrete Pipe—An Update*, Corrosion 98, Paper No 637, NACE International, Houston, TX (1998).
37. *The Thaumasite Form of Sulphate Attack: Risks, Diagnosis, Remedial Works and Guidance on New Construction*, Report of the Thaumasite Expert Group for the Department of the Environment, Transport and the Regions, London, UK (1999).
38. Howell, K. M., Corrosion of Reinforced Concrete Slab Foundations, *Materials Performance*, 29, 12 (1990): p. 15.
39. Das, S. C., *A Novel Approach to Stop Reinforcement of Subterranean Structures by Cathodic Protection*, 5th International Conference on Structural Faults and Repairs, Edinburgh, UK (June 1993).
40. Chadwick, R., Chaudhary, R., Electrochemical Protection of Buried Reinforced Concrete Foundations, *Corrosion Management*, 22 (February/March 1990): p. 6.
41. NACE RP 0290-2000, *Impressed Current Cathodic Protection of Reinforcing Steel in Atmospherically Exposed Concrete Structures*, NACE International, Houston, TX (2000).
42. BS EN ISO 12696: 2000, *Cathodic Protection of Steel in Concrete*, Comité Européen de Normalisation (CEN), Brussels, Belgium (2000).
43. Wyatt, B. S., *Trends in the Design and Management of Cathodic Protection Systems for Pipelines and Storage Tanks*, Conference on Materials Selection and Application in the Chemical Process Industry, London, UK (November 1998).

44. BS EN 12954: 2001, *Cathodic Protection of Buried or Immersed Metallic Structures—General Principles and Application for Pipelines*, Comité Européen de Normalisation (CEN), Brussels, Belgium (2001).

45. NACE RP 0169-2002, *Control of External Corrosion on Underground or Submerged Metallic Piping Systems*, NACE International, Houston, TX (2002).

46. Gummow, R. A., Cathodic Protection Potential Criterion for Underground Steel Structures, *Materials Performance*, 32, 11 (1993): p. 21.

47. Dearing, D. M., The 100-mV Polarization Criterion, *Materials Performance*, 33, 9 (1994): p. 23.

48. ISO/FDIS 15589-1, *Petroleum & Gas Industries—Cathodic Protection of Pipeline Transportation Systems—Part 1: On-land Pipelines*, International Organisation for Standardisation (ISO), Geneva, Switzerland (2003).

49. NACE RP 0100-2004, *Cathodic Protection of Prestressed Concrete Cylinder Pipe*, NACE International, Houston, TX (2004).

50. Unz, M., Cathodic Protection of Prestressed Concrete, *Corrosion*, 16, 6 (1960): pp. 289–297.

51. Spector, D., Pre-stressed Reinforced Pipes—Deep Well Groundbeds for Cathodic Protection, *Corrosion Technology*, 9, 10 (1962): pp. 257–262.

52. Deskins, R. L., Cathodic Protection of a Mortar Coated Steel Water Distribution System, *Materials Protection*, 5, 9 (1966): p. 35.

53. Deskins, R. L., Cathodic Protection Requirements for Concrete Pipes, *Materials Performance*, 17, 7 (1978): p. 50.

54. NF A 05-611, *Protection Cathodique des Armatures du Béton—Ouvrages Enterrés et Immergés* [*Cathodic Protection of Reinforcing Steel in Concrete-Bureid and Immersed Structures*], Paris, France: Association Française de Normalisation (AFNOR), (1992).

55. Peris, M. G., Guillen, M. A., Cathodic Protection to Existing Prestressed Concrete Pipelines, *Materials Performance*, 34, 1 (1995): p. 25.

56. Benedict, R. L., Ott II, J. G., Marshall, D. H., White, D., Cathodic Protection of Prestressed Concrete Cylinder Pipes Utilizing Zinc Anodes, *Materials Performance*, 36, 5 (1997): p. 12.

57. Garman, F. A., Al-Maadani, F., *Experience of using Zinc to Prevent Corrosion of Pre-stressed Concrete Cylinder Pipes in High Resistivity Soils*, 16th International Corrosion Congress, Beijing, China (September 2005).

Chapter 7

Design of a cathodic protection system for exposed reinforced concrete structures

Paul M. Chess

CONTENTS

7.1 INTRODUCTION

Cathodic protection (CP) design of 'conventional' steel structures in soil or water is a well-established discipline that involves an estimate of the size and geometry of the structure to be protected, current requirement calculations and a design of the most suitable type and size of groundbed.

The design of a CP system for reinforced concrete is not as well documented as for 'conventional' CP systems, but in compensation, there are some variables in the design of an underground or undersea system, which are relatively fixed for protecting steel in concrete aboveground level.

The purpose of this section is to discuss the various factors that the designer should consider and give some information to provide a satisfactory CP design. A large document called Concrete Society Technical Report

(CSTR) No 73 (2011) has recently been published and this chapter should be considered as additional in its information [1].

7.2 SYSTEM DESIGN

The two most important factors for the designer of a CP system for steel in concrete to consider are the current density required on the steel and the current distribution path, that is, the steel reinforcement where protection is required. Beyond these requirements, the designer has a myriad of other concerns such as cost, aesthetics, weight, durability, life expectancy, maintainability and track record, to name but a few. These secondary factors may often conflict and the correct solution will normally be a compromise of a commercially available anode and other materials that most satisfactorily resolve the problems. On rare occasions, no commercially available anode is available and the designer might have to utilise a hybrid or experimental anode. In contrast to the juggling of the secondary considerations, the first two factors, that is, current density and distribution, should not be compromised as the only purpose of the system is to reduce or prevent corrosion of the steel reinforcement, and if there is insufficient current density or inadequate current distribution, this objective will not be achieved.

7.3 OVERALL SYSTEM PHILOSOPHY

When appraising a structure, an assessment of which steel is considered most at risk and the area is most desired to protect should be made. For example, on beams close to the sea, the worst damage was on the outer areas exposed to the prevailing wind, and the areas that the structural engineers were most worried about ongoing corrosion were the bottom outer layers of steel, so the biggest concentration of CP current was made in the bottom outer area. Other factors that should be considered at the outset are as follows: the life expectancy of the CP system, likely future maintenance, what any other refurbishment is going to be undertaken and probably most fundamental to the client—budget.

Generally, a CP system should be designed so that the corrosion rate is minimised for the design life of the system. Occasionally, if there is only a limited life requirement, the anode area or quantity can be reduced. Any other refurbishment of the structure is also of large importance in determining what anode system is selected, that is, is localised strengthening going to be used? If this is the case, is the repair area going to be broken out to below the first level of reinforcement steel? How are the repairs going to be made?

Any of the preceding factors can have a profound influence on the type and design of the CP system and should be determined as soon as practicable.

7.4 CURRENT DENSITY REQUIREMENT

The selection of a suitable current density output is critical for the CP designer. Unfortunately, there was until recently little information on national or international standards to help. Indeed, some early publications were misleading in that they implied that a fixed current density is sufficient to provide CP in all circumstances.

In recent documents such as the Concrete Society Technical Report No 73 (2011), a link between the exposure conditions and current requirement is made. The author's practical experience has shown that the current density requirement is extremely dependent on the steel's corrosion state before CP is applied, which is generally related to the environment surrounding the steel.

For example, if the concrete surrounding the steel is alkaline, there is little chloride present, the diffusion rate is very low and the steel is not actively corroding, a very low current density will be sufficient to prevent any corrosion occurring in the future. At the opposite extreme, areas with minimal concrete cover, a warm, wet, fluctuating environment with high oxygen and chloride levels will have a very high current density requirement. An example of this is a sea water intake in the Arabian Gulf. Often a hundred times greater current density is required on this structure than the first example to reduce the corrosion rate to the same level.

A practical guide, from the author's experience, is given in Table 7.1 to achieve about a two-decade reduction in corrosion activity (99%). It should be noted, however, that the most accurate and effective way of defining the required current density is to undertake a CP trial as discussed later in Section 7.6.

Table 7.1 Practical CP current density requirements for varying steel conditions

Environment surrounding steel reinforcement	Current density (mA/m² of reinforcement)
Alkaline, no corrosion occurring, low oxygen resupply	0.1
Alkaline, no corrosion occurring, exposed structure	1–3
Alkaline, chloride present, dry, good quality concrete, high cover, light corrosion observed on rebar	3–7
Chloride present, wet, poor-quality concrete medium-low cover, widespread pitting and general corrosion on steel	8–20
High chloride levels, wet fluctuating environment, high oxygen level, hot, severe corrosion on steel, low cover. This amount of current is only required in exceptional circumstances	30–50

7.5 CURRENT DISTRIBUTION

Of equal or perhaps greater importance than the total current density applied is the way that it is distributed. The optimum current distribution requirement should be assessed from the steel reinforcement arrangement, the extent of corrosion spread and the level of activity.

As part of a CP survey, the areas of active corrosion should be defined. Normally, the highest level of current should be injected at these locations. On the contrary, if extensive concrete repairs are to be carried out due to cracking, delamination and spalling of the concrete cover, then these areas may become passive and have a lower current demand than the less-damaged parts of the structure.

The 'localised' current distribution is very dependent on the anode type and even more importantly on variations in the concrete resistivity. When there are limited changes in resistivity of the concrete, surface-mounted anodes such as meshes and conductive coatings give an even, lateral distribution from the surface while discrete anodes embedded into the concrete give a spheroidal, or sometimes 'rugby ball—shaped' distribution around the central axis of the anode rod. This latter system can be made to achieve a relatively even lateral distribution, if sufficient anodes are used.

Several efforts to mathematically describe the current distribution from the anodes have been made as this has proven effective for traditional CP; however, this has not been validated for steel in concrete and so is rarely used. It is difficult to describe, in mathematical terms, the current distribution in reinforced concrete. This is due to the large changes that occur; firstly, in the resistivity of the concrete, secondly, in the resistivity of the steel to concrete interface as current is passed and thirdly, the profound effect of orientation and density of the steel reinforcement. However, as it is very important for a CP designer to know where the protection current is likely to spread, some examples of typical distributions are discussed:

1. In a simple slab with touch dry concrete and a laterally uniform chloride penetration from the outside to a depth of 70 mm into the concrete, and where there is 50 mm of cover and a second layer of steel at 300 mm depth, the following current distribution can be anticipated. With an anode uniformly spaced on the top of the concrete, a reasonable area to design on is the steel top surface mat surface area multiplied by 1.5. This multiplication factor takes into account links and tie wire on the top mat. Due to the limited chloride penetration, it is unnecessary to allow sufficient output to cathodically protect the lower reinforcement. The lower mat is only to be considered as a current drain and about 10% of the total current applied may be expected to reach here. This drain is relatively low because of the absence of chloride, high cover and limited oxygen resupply. This

area is then multiplied by the current density anticipated and the area to give the current required for the zone.

2. In a simple slab as in (1) where there are also a substantial amount of shrinkage cracks, it is likely that chloride has penetrated deeper into the slab at the cracks. As there is, in effect, a lower cover depth and a higher oxygen availability at these locations, a higher current density is required to prevent further corrosion of the steel near the cracks. In these areas, twice the current output may be needed. This can be achieved by doubling the output of the anode system over the entire slab, or more economically achieved by engineering an increase in the output of the anodes in these localised areas.

   The ability to increase the current output in a localised area depends on the anode type. Coated titanium mesh output can be increased by welding a second layer to the original mesh or by using a thicker anode mesh in the localised area. Conductive coatings can have more primary feeders installed at these locations, the conductive layer applied more thickly or the surface roughness increased to enlarge the anode area. Discrete anodes can be increased in size or more can be installed in the same area. Sometimes, it may also be possible to apply additional CP anodes on other faces of the structure to protect these particular locations or even use embedded anodes in concert with surface-applied anodes.

3. Where access to apply the CP anode on a structure is limited and yet there are several layers of reinforced steel with the concrete contaminated with chlorides, the designer has severe problems. One example of this type of structure is an immersed tunnel where chloride has permeated in from the outside but the oxygen flow is from the inside out.

   In this case, an inadequately designed CP system could fail to stop all the corrosion occurring and may move the anodic sites deeper into the concrete. This may not always be a problem as the corrosion rate will be significantly lower in this area due to the low oxygen availability. If CP is still considered suitable despite these caveats, then the current distribution requirement may be based on corrosion prevention of the innermost layers of reinforcement steel only. In this example, there will still be a substantial current drain to the outer layers of steel. Thus, allowance should be made for protecting at least two and a half times the area of the innermost steel layer to provide sufficient current density for the protection of the most at risk (the inner layer) steel.

   It should be evident from these examples that it is difficult to generalise on the current distribution in more complicated situations. It is thus recommended that, when the structure is different to those which have been protected previously, a trial is undertaken during the CP design survey to enable an assessment of current distribution.

7.6 TRIALS AND TESTING

It is apparent that there are several significant factors that impact on the current output required from the anode, as discussed in Sections 7.4 and 7.5. To minimise the likelihood of overdesign or underdesign, it is good practice to install a trial on the structure at pertinent points during the design survey when dealing with complex structures or unusual conditions. The trial system should typically comprise at least 1 m² of concrete surface area. The minimum size of the trial is dependent on the amount of steel and the resistivity of the concrete. The reason for this is that with low-resistance concrete and dense steel reinforcement, there will be a large current drain to outside the protected area, which could lead to a substantial reduction in potential changes recorded on the steel, that is, a substantial underestimate of the effectiveness of the trial CP system. For instance, trials on the example in Section 7.4 (point 1) need to be a minimum of 1 m² in area. It may be necessary to trial a 10 m² area in Section 7.4 (point 3) to reduce this effect. A power supply (dry cell battery or car battery is normally sufficient) and a negative connection are required to complete the circuit. A portable reference electrode can be used for measuring potential changes on the steel reinforcement. The anodes normally used for such trials are either conductive coatings, discrete anodes or ribbon, due to their ease and speed of installation (mesh and overlay and the like require a significant amount of plant to install).

Before powering up the trial system, the ideal is to take several surface half-cell measurements, in and around the protected area, and construct an iso-potential map. After energising, the system makes another iso-potential map representation of the structure. The changes in potentials over the trial area can then be calculated. The aim of a successful trial is to demonstrate that all areas are now net cathodes and that particular potential criteria, notably potential 'shift', are being met.

It is normally most satisfactory to initially run the trial at the 'best guess' voltage level considered necessary for protection and measure the current as the trial polarises. If possible, the system should be operated for a protracted period (at least a week) at the same output. Unfortunately, this idea is frequently impractical and valuable information on the current density requirement and current distribution can be obtained in a day. It should be noted that there is a substantial current drain to the edges of a trial that will not occur with the main installation.

7.7 ZONES

For the CP system to be effective continuously, individual areas where there is a significant change in the environment of the steel reinforcement should be protected by separate CP circuits, that is, separate CP zones. These changes are normally discerned in the CP survey by large variations in the resistance of the concrete and potential of the steel. These can be caused by changes in moisture content, chloride contamination, cover or geometry of the component in a structure.

When using an anode system where only a limited amount of current increase can be imparted at specific areas, or the anode type is prone to large changes in resistance in accordance with environmental factors, that is, wetting and drying, provision should be made for an increased number of zones.

Typically, zones of the order of 50–100 m² are recommended, but this is dependent on the structures' form and environment. For example, in selecting zones on a marine structure, as shown in Figure 7.1, it is common to split the structure into separate zones relative to the water level. If the areas at the individual level are small, as in this example, it is normal practice to electrically connect the anode areas together even if they are physically separated on the structure. Care should be taken, if this is done, that the steel reinforcement to these individual areas is electrically continuous. It is important to recognise that it is the anode that is split into zones. The reinforcement may or may not be originally continuous, as shown in Figure 7.1. All steel in cathodically protected areas should be connected back to the negative terminal of the zone power supply.

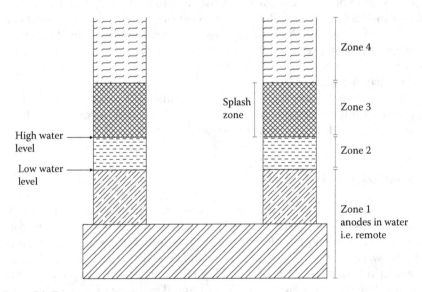

Figure 7.1 Zoning on a marine support structure.

7.8 SPECIFICATION

The specifier has a considerable responsibility for ensuring the quality and performance of the installation and should carefully consider in tandem with the client regarding what will be the most cost-effective approach to the project.

There are several approaches that can be taken, which range from the specifier designing the system, specifying all materials and quantities and passing to a term 'contractor' while at the other end of the spectrum having a performance specification and allowing the tenderers to subcontract out the design, materials and installation.

Typically, with a performance specification, the contractor gets to choose the anode type, zoning and cabling among other factors. This performance part of the specification normally gives the current density required, life expectancy, details of the aesthetics and the electrochemical performance in meeting the CP criteria. From practical experience, if this approach is followed, then the track record for the contractors put on the tender list should be closely examined and controlled. This approach is likely to lead to the cheapest tender prices but will probably compromise on quality.

The most common option is a performance specification with an outline design specifying but not naming the products desired. This has the advantages that the consultant generally gets the products desired while keeping liability with the contractor. It is a reasonably good system but can be abused when contractors try to cut costs by substituting cheaper materials.

In Denmark, after initially following the heavily specified route, it has recently become common for the consultant to specify by manufacturers the major cost items and allow a limited list of specialist subcontractors to be allowed to bid for the project. This has been found to provide a high-quality system at a reasonable price. A reason for this is that it necessitates a greater cooperation between the parties while avoiding the paperwork of formal partnership.

The Concrete Society has published a Technical Report No 73, which is a guidance document and specification for CP of steel in concrete, which gives a detailed description of the anodes.

7.9 ANODE SELECTION

This is the critical choice for the structure and the various commercially available anodes, with their advantages and drawbacks outlined in the following discussion. But firstly, it is worth considering that anode is the interface between electrically passed current and ionically passed current. Due to the relatively thin levels of cover and high resistance nature of the concrete, the anode has to be placed over a substantial proportion of the concrete surface area, so the current density output of the anode is necessarily much lower than a traditional anode. At the electronic to ionic interface, there are chemical reactions that tend to generate acid and often chlorine gas. These chemical reactions cause the breakdown of the adjacent cement paste with implications for looks (the cement paste normally contains bound iron that can give red staining), adhesion (if the anode is 'glued' to this cement surface, the bond strength can be reduced) and passage of current (as the cement reacts, the water-soluble products can be washed away leaving voids that can increase the resistance to the passage of electricity). These deleterious effects are normally limited by having a specified maximum current density. This is a sensible approach but great care in design needs to be taken that certain areas in the zone are not receiving the vast bulk of the current and other parts of the zone receiving virtually nothing.

The anode types for impressed current use in common commercial service are as follows:

1. *Conductive coating carbon-loaded organic coatings:* These are paints that are normally derived from outdoor technical paints that are loaded with carbon. In the earlier versions of the paint, they were normally solvent-based but more recently they have been water-based. These anodes are fed by primary anodes, which are now normally uncoated titanium strips glued to the original concretes surface. These anodes have proven to be inexpensive and relatively simple to install, but if inadequately designed and installed can have poor current distribution. They have also proven susceptible in harsh environments to failure. This anode type has a declining market share with the development of a wider range of impressed current anodes.

2. *Sprayed metal coatings:* Several of these thermally sprayed metals have been used, namely zinc, aluminium, stainless steel and titanium. Normally, titanium feeder strips or pads glued to the surface are used as primary anode feeders. The most common anodes presently in use are zinc and an alloy of zinc and aluminium, although these alloys are more usually applied as galvanic anodes. These anodes have an advantage over a conductive paint in that as there is acid build-up at the anode to concrete, this oxidises the metal and provides a strong bond keeping the anode adherent. This oxidisation also tends to electrically isolate the anode so often that ionic solutions are placed on the anode to improve the situation.

3. *Modified mortar:* There is one maker of this type of anode that is composed of a cementitious mortar with nickel-coated carbon fibre threads in the compound.

   This is fed by a coated or uncoated titanium strip affixed to the concrete surface. The gap between the primary anode feeders is typically about 600 mm. The conductive mortar is sprayed onto the surface-prepared concrete.

The system has shown itself to be durable in benign and moderately harsh environments and has been moderately successful in the world market.

4. *Mixed metal oxide (MMO)—coated titanium ribbon placed before casting*: MMO is a ceramic coating that is electrically conductive; it is composed of ruthenium and tantalum oxides. This is sprayed and baked onto a titanium substrate. Titanium is used because below 8 V it has a passive oxide layer that stops it conducting, so all the current is passed out through the MMO. This anode combination is very successful and has become the primary anode of choice for all CP of concrete applications. This type of anode in the form of a perforated expanded ribbon is commonly used for the protection of new structures. The anode is typically spaced off the steel reinforcement. Earlier, this was done using cable ties and plasterboard, but on more recent installations, purpose-made clips have been used. This form of future corrosion prevention has become popular in certain countries such as Saudi Arabia where there is a history of severe corrosion problems. The biggest concern for these installations has to be the decision on where the future corrosion is likely to occur and zone the ribbons appropriately. The major issue during installation is ensuring electrical isolation of the anode from the steel throughout the concrete casting process.

5. *Coated titanium ribbon in slots in the original concrete*: At present, this is possibly the cheapest anode type that is in widespread use with several variations on placement. The most common is cutting vertical slots with an angle grinder into the concrete surface and cementing with mortar. Slots are typically 300–450 mm apart. Other variations are making a wider shallow slot and placing the anode sideways and finally putting the ribbon on the surface and spraying mortar over it. The biggest problem with this system is the variation in output along the length of the ribbon. This means in practice that you can get very high current outputs in localised areas that will break down the mortar and leave voiding around the ribbon. In marine or similar environments, this voiding will then be filled with water that will have a much lower resistance than the mortar and so even more excess current will be delivered at these locations and virtually nothing at other parts of the ribbon.

The only way to minimise this effect is to use a lot of zones, plan where these zones may be most effective and promote water run-off from the structure.

This anode has the advantage of not changing the profile or loading of the structure. It requires sufficient cover to the steel to avoid the risk of short circuits between anode and steel. It leaves stripes on the surface.

6. *Coated titanium mesh covered with concrete or similar:* This anode is created when a mesh looking like chicken wire is unrolled and secured to the prepared concrete surface. This is then covered with a cementitious material most commonly sprayed concrete. This system has been in use for many years and has an excellent current distribution at low driving voltage. In most situations, it has proven to be a durable and an effective anode. It is fairly expensive and puts weight on a structure. In some cases, there has been premature failure that is normally caused by a breakdown between the original concrete and the new cementitious material. In some instances, it has been conjectured that this has been exacerbated by the passage of current from the anode moving water through this interface.

7. *Drilled in discrete anodes:* These anodes are placed in holes drilled in the original concrete. The first discrete anodes were manufactured and installed more than 25 years ago. The purpose was to make hot spot CP of an especially aggressive exposed area. The original anode used a coated titanium rod as primary anode. The electrical connection to the concrete was established with a graphite paste. As feeder wire, a titanium wire was welded to the primary anode. Many of these installations are still successfully operating and this gives the discrete anode a real-life durability record that is unparalleled in Europe.

Discrete anodes have been further developed so that an optimal current output and distribution is established using a built-in resistor. A mechanical connection system of the feeder wire ensures a quick and durable connection. The connection between the anode material and the concrete is commonly made using a hydraulic cementitious grout.

Using a fine expanded MMO-coated titanium mesh, the electrical conductance between primary anode and backfill is improved compared to a massive rod. Also, as there is a cementitious mortar on the inside of the tube, the amount of buffering capacity of the grout is effectively doubled.

Discrete anode systems can be installed using few tools and do not increase the weight and dimension of the structure. This is of particular importance in car parks where the height may be very limited and on some type of bridges.

Anodes are typically spaced at 350–450-mm centres depending on the steel surface area, anode size and likely current requirement. Drilled in anodes can be difficult to place when there are large amounts and randomly placed reinforcement in a structure. Drilled in anodes are normally not cost-effective when there are thin sections of structure.

7.10 CABLES AND CONDUITS

The size of the cable required for the anode and cathode wires is now normally calculated so that the voltage drop is less than 5% along the total run of the cables.

For an atmospherically exposed reinforced concrete structure, the cabling should normally be run in conduit or buried in the structure at all locations. If thorough quality control on-site is adhered (necessary whatever cable dimensions are selected), then the most logical approach is to use cable cores with the minimum size necessary to pass the maximum design current.

The voltage drop of the circuit is calculated from the cable resistance for the selected conductor cross section, multiplied by the cable length, taking into account the required driving voltage of the anode. As a rule of thumb, for both DC positive and DC negative feeder cables, the total voltage drop should always be less than 2 V. The driving voltage required by the anode system depends on the ease with which it passes the current onto the steel reinforcement. This is dependent on the concrete resistivity and other factors; however, a rough guide is given in Table 7.2. This table gives the required voltage to pass 10 mA/m^2 of steel (which is a fairly high level of protection) after a few years of operation of the CP system on a surface dry concrete that has some chloride contamination.

Normally, a few years after a CP system is energised, the current demand required to prevent corrosion of the reinforced steel is significantly reduced. At this point, the current density the system is delivering should be lowered by reducing the driving voltage. Looking at Table 7.2, it should be apparent that the wattage required to pass the same amount of current could be more than doubled between one anode type and another. On a large system, this will have a significant effect on the system running cost over its life.

Table 7.2 Typical driving voltages required by different anode types

Anode type	Typical driving voltage required after 2 years (V)
Mesh and gunite	4
Ribbon in slot	6
Solvent-based conductive coating	8
Water-based conductive coating	8–15
Discrete anode	6 (2 V of which is pre-resistor)
Sprayed zinc	>15
Conductive mortar proprietary system	8

The most suitable cables for use in concrete CP installations vary according to the wires' function and the environment.

> *Reference electrode.* This is passing a very small amount of current but the sheathing is required to be directly buried in the concrete for the lifetime of the installation. The chief requirement of the sheath is that it is alkali-resistant over a protracted time and for this either cross-linked polyethene (XLPE) or a fluorocarbon is most suited. Normally, the conductor is copper cored. For an extended lifetime, stainless steel or titanium can be considered. Stainless is cheaper and can be more flexible, that is, it can be procured in a stranded form.
>
> *Anode feeder wire.* If the anode feeder wire, ribbon or rod is to be buried in concrete, then it should be made using titanium with connection to copper-cored wires made on the surface of the structure, preferably in junction boxes. This is because the titanium to MMO-coated titanium connection (the MMO-coated titanium is used in most anode types) can be made without having to devise a completely impermeable encapsulant. This connection detail is so critical because at any point where there is a passage of current there is likely to be acidity and an electrochemical oxidation of the conductor. If the titanium anode-free wire has a sheath, which is useful to prevent unintended shorting, the chief requirements are that it is alkali- and acid-resistant over a protracted time and for this either XLPE or a fluorocarbon is most suited.
>
> *Cathode (steel reinforcement) return wire.* The return cathode wire is protected in some part by the CP and is normally replicated many times over (the structure is normally electrically continuous) so it is acceptable to use a copper-cored wire. The chief requirement of the sheath is that it is alkali-resistant over a protracted time and for this either XLPE or a fluorocarbon polymer is most suited. To provide additional mechanical protection, it is common to specify an inner sheath and an outer sheath.

Positive connection damage is one of the most common failure modes in a CP system. Cable failures are also common. These can normally be attributed to mechanical damage during and after installation; having unprotected cable in the structure, that is, run without conduits; overtensioning in installation, that is, stretched cables that are then thermally cycled and, finally, bio-interference such as marine attack. Other common failures encountered are at line splices that are often woefully underspecified. Some specifications forbid the splicing of cables and also require factory (rather than site)-made connections to avoid problems. There are also problems at junction boxes. These are often located in areas where they are liable to be flooded or suffer from water ingress due to an inadequate IP rating of the box or glanding.

Most of these problems can be prevented by good design and site practice and following the guidance provided in BS EN ISO 12696 and CSTR73 [2]. As contractors try to use multicore whenever possible, for cost-saving reasons, it is normally impractical for the designer to specify glanding sizes and numbers on the junction boxes and power supplies, but they can specify that they are sited in as dry a location as possible and the glanding is all facing downwards. Site supervision should ensure that the glands are of correct size for the cable. When in doubt about the durability of junction boxes in exposed areas, they should be filled with a non-acid, petroleum jelly to preserve the integrity of the connections. If in the event that flushing is considered to be possible, then a proper potting compound/encapsulant should be used.

It is good practice to separate the DC positive and DC negative feeder cables as much as possible, preferably putting them into separate junction boxes, to prevent the possibility of galvanic corrosion (if there is even a small amount of moisture, you can get a short circuit between the positive and negative cables, which causes destruction of the positive cables). When inline splices are unavoidable, the joints can be made with a mechanical sleeve splice, faired with a suitable mastic epoxy putty and at least one mastic heat shrink sleeve and preferably two for anode feeder connections. The use of the appropriate waterproof jointing kit such as the 3M Scotchcast system is also recommended. All terminations and joints are better fabricated in the factory than on-site.

For cables that are out of the concrete, in most cases, they should be run in conduit and trunking to the power supply cabinets. There are several factors to consider before deciding on the most appropriate type. These are the corrosivity of the surrounding area, vandalism, fire restrictions and aesthetics. In some of the early installations, galvanised trunking was used outdoors and this generally now looks terrible and has often corroded to perforation. This sends a dreadful message to a client, that is, on an anti-corrosion system there is rampant corrosion.

If metallic trunking is used, this typically will be stainless steel of the appropriate grade for the environment. Plastic conduits have proven to be durable but the cheapest, that is, polyvinyl chloride (PVC), has fire limitations and should be avoided for internal installations. There is widespread availability of high-density polyethene (HDPE) or XLPE, Both are thermosoftening polymers, so these should not be used in carrying any load under direct sunlight. Fixings should be to the same standard as the conduit, and if crossing a CP zone, it should be non-metallic or stainless steel in a resin anchor fixing.

7.11 REFERENCE ELECTRODES AND OTHER MEASURING DEVICES

There are several types of reference electrodes that are commercially available and suitable for burial in concrete. Due to the relatively poor performance of reference electrodes over protracted periods in the early days of applying CP to reinforced concrete, a significant amount of work has been undertaken on assessing how reliable the particular types are.

Reference electrodes can be categorised into two types for burial into concrete, namely true half cells and 'inert' or 'pseudo' reference electrodes.

True half cells can be defined as an element in a stable and reproducible dynamic balance with its ions. For example, the silver/silver chloride reference electrode has a silver rod or mesh coated with silver chloride in the middle of the unit with a saturated electrolyte of silver or potassium chloride solution that is normally made into a gel to prevent or slow down leakage. This gel forms the electrolytic connection between the concrete with the interface made through a porous plug either on the flat face or in some designs over all the cylinder shape. The other commonly used true half cell is the manganese dioxide electrode in a stainless steel housing with an alkaline gel. Reference electrodes for use in concrete are 'double junction' electrodes that minimise contamination or dilution of the electrolyte through which the ions flow.

Inert electrodes are units where the active element has an extremely small dynamic equilibrium coefficient between the element and its ions in concrete. Graphite, platinum or mixed metal oxide—coated titanium have been found to be effectively inert in concrete, whether it contains chloride or not, and thus maintain a relatively stable potential. As these three elements can withstand some anodic discharge, there is little material loss when a potential measurement circuit is left open even for a period of time. This means that they are particularly suited to automatic control systems where readings can be taken on an hourly basis. The major limitation is that their potential varies with the oxygen level at the electrodes surface, so their potential fluctuates with, for example, concrete moisture level.

If the CP system is operated using 4- and 24-hour 'decay' criteria, then inert electrodes are perfectly satisfactory and have the advantage of offering greater robustness and a longer theoretical life as long as the water saturation level does not fluctuate significantly over this time, for example, in a tidal zone. If the system is operated using absolute potential criteria or has the possibility of reaching very negative potentials, then true half-cell reference electrodes are required. The true half-cell type of reference electrode should also be specified if comparisons between the steel reinforcement potential before energisation with free corrosion criteria and after CP are made.

The major advantage of the inert reference electrode/potential decay probe is its indefinite life. Therefore, on very long-life systems, where it will not be possible to replace the electrodes, this type should be chosen. The most satisfactory arrangement is to specify a mixture of true half cells and inert reference electrodes for a single structure that is placed in pairs with the true half cell used for direct measurements and the inert electrodes, checking the calibration of the half cells and taking over if the true reference electrodes become unstable.

The most commonly used reference electrodes and their categories are given in Table 7.3.

It is essential to obtain reference electrodes from reputable manufacturers with an established track history of providing electrodes for embedding in concrete.

Assuming that a reputable brand has been specified, the most important practical factor determining the satisfactory performance of the reference electrodes is the integrity of the interface between the reference electrode and the concrete into which it is embedded. If there are voids in the electrolytic contact, and these dry out after the cement cures, the resistance of the circuit increases and new electrical pathways may occur so that spurious readings result. To minimise this possibility, the interface area, that is, the size of the porous plug for a true half cell or exposed element for an inert cell, of the reference electrode specified should be maximised. On some of the latest monitoring systems, the impedance of the reference electrode circuit is measured at the same time as the potential and this gives a direct assessment of the state of health of the reference electrode.

The way in which the reference electrode is connected electrolytically to the original concrete is very important. The most common way is to use a 'clean' (no added chlorides) proprietary mortar with minimal anti-shrink agents. An alternative favoured in some Middle Eastern countries is to pre-pot the reference electrode within a small diameter concrete cylinder in the workshop and then further encapsulation on-site.

Table 7.3 Common electrodes specified for burial in concrete

Type	Abbreviation	Potential compared to CSE (mV)	Category
Silver/silver chloride/potassium chloride (0.5 M)	Ag/AgCl/KCl	+70	True half cell
Manganese/manganese dioxide	Mn/MnO$_2$	+95	True half cell
Graphite	Gr or C	−50	Inert
MMO-coated titanium	MMO(Ti)	+110	Inert
Platinum-coated titanium	Pt(Ti)	+100	Inert

Note: CSE, copper/copper sulphate electrode; MMO, mixed metal oxide.

The location of the reference electrode is of great importance, as this has a large influence on the extent and location of the steel reinforcement measured by the reference electrode. In early practice in the United Kingdom, where a large amount of reference electrodes were put into relatively small volumes of concrete, it was considered reasonable to strap the reference electrodes to the steel reinforcement; however, this had the disadvantage of putting the steel that has the greatest potential influence in new fresh mortar and restricting the area of the reference electrode readings. Unfortunately, this placement technique of reference electrodes is still prevalent, whereas the correct approach, as stated in BS EN ISO 12696 [2], is to replace the reference electrodes without disturbing the original concrete around the steel so that the potential measured is that of steel in contaminated concrete and not in pristine repair material.

Monitoring connections to the steel reinforcement near the reference electrodes are commonly specified to minimise the error from the flow of current in the steel reinforcement when being cathodically protected. It is reasonable but not vital to have a connection to the steel for each reference electrode in a zone if the steel is continuous. If there is only a small amount of current flowing (typically of the order of tens of milliamps) through the CP circuit, the DC negative power return for the CP system can be used against the reference electrode with little errors accruing.

When the reference electrodes are installed, it is of some value to designate an area on the surface near their individual location and determine their potential against a calibrated portable reference electrode placed on the surface. If substantial drifting of the potential occurs, then the internal reference electrodes should be replaced or ignored. This is sometimes made difficult to impossible by the type of anode being used.

Reference electrodes are by far the most common method of determining the effectiveness of CP; however, other methods have been used in the past. One example is where a section of reinforcement steel is cut and electrically isolated still within its original concrete, and the current flow between the steel and the remainder of the reinforcement steel case is measured. The concept is that the electrically isolated section of reinforcement steel that is corroding would be a net current provider to the remainder of the steel reinforcement in the structure. As the CP system is energised, the current flow between the isolated section and the remainder of the steel reinforcement would reduce and eventually it would become a net receiver of current. At this point, corrosion activity would be stopped as the previously anodic (corroding) areas within the isolated bar would become cathodic. This information would then imply that other similar areas had received sufficient current and could be used to set up the system.

A similar concept has been used for 'current pickup probes' where a drilled hole was made in the concrete, a steel bar was inserted and grouted into place using a mortar having a higher chloride level than the original concrete. The steel bar was electrically connected to the structure with a low-value resistor and the corrosion current was measured as the CP system was energised progressively. When the bar became cathodic, the sufficient current was deemed to be provided to ensure protection of all the rest of the reinforcement. This has not been used for several years.

Other measurement systems have been tested in a laboratory and field trials but do not appear to be in widespread use on commercial CP systems as yet. The increased reliability of reference electrodes and more confidence in the potential criteria for the control, a CP system has meant that alternative measurement techniques have become less relevant.

7.12 INTERACTION

When designing the CP system, attention must be given to the possibility of interaction with other components. The most problematic forms of interaction are large DC currents and these, typically, can be caused by electrically generated train or tram traction systems often found at the ground/air interface. There is also often a significant amount of electrical 'noise', particularly at 50-Hz frequency, where there is grounding from nearby electrical apparatus. This occurs quite often on marine structures. Reference electrodes in the tidal zone can pick up this electrical 'noise'. AC 'noise' is not normally a problem from a protection point of view but can give problems with readings from reference electrodes if the cables run in parallel with unshielded AC cables. This can often be overcome by understanding the likelihood problem and using suitable electronic filtering or cable screening.

If interaction problems are encountered, the solution is normally the same as in a 'traditional' system. These include bonding the nearest part of the interaction circuit to the system cathode, either directly or by using resistors or diodes. Another approach is to put sacrificial anodes connected to the reinforcement into the electrolyte and use them as preferential current receivers. Sometimes, cables, particularly those carrying reference electrodes potentials, can pick up induced currents and when this occurs screened cables are required.

7.13 CONTINUITY AND NEGATIVE CONNECTIONS

Unlike 'conventional' CP systems for pipes or other buried metallic components, electrical integrity of the cathode for a steel-reinforced concrete structure is more difficult to prove and cannot be checked completely, as removal of all the concrete cover is not practical and often precisely the reason for installation of the CP system. Therefore, an integral part of the design is to confirm the electrical continuity of the steel reinforcement. In the design survey, this can and should be tested directly using the techniques described in BS EN 12698 [2].

In most cases, there will be excellent electrical continuity between the reinforcing bars as they will have been securely connected together to form a structurally robust cage or mat during the construction process. However, there can be problems on older lightly reinforced structures and severely corroded structures. Assuming that the steel has been conventionally tied with wire, the most critical factor is to estimate or determine the amount of corrosion that has occurred between the rebar and tie wire. If there is a significant amount of corrosion, then the current that can be passed at this connection may only amount to a few milliamps and either additional reinforcement continuity bonding will be required, or a large number of individual DC negative connections need to be made. As a guide, a DC negative for every 50 m^2 should be provided with this number quadrupled if any concerns on the likelihood of poor continuity are expressed. However, there should always be at least two negative connections for electrical redundancy.

When designing a CP system, particular care should be taken to ensure that there is electrical continuity across expansion joints, dry joints or other discontinuities, such as where there are different concrete colours that are indicative of separate pours.

In general, it is recommended that the entire DC negative system is made electrically common. For example, when there are precast components that are electrically discontinuous, these should be electrically commoned. Where a structure is made of precast units, these may be connected by dowels with no continuous steel connection. In this case, the various units and the dowels all need to be made electrically continuous.

Particular examples where it may not be pertinent to electrically common a structure are rare and normally involve CP systems where there is concern that the current distribution will be excessive or variable, or there is possibility of stray current interaction. An example of the former is individual piers for a marine bridge where the deck is isolated from the substructure by the bearings. In this case, each pier can be protected with their own DC negative and DC positive circuits. When this is done, the communication wire should be of the fibre optic type or a radio data communication system should be used. When designing a CP system for a tunnel, which has individual precast segments and an electrical traction system that is ground earthed, it is likely that if the segments are electrically commoned, then substantial interaction stray currents would be induced onto the reinforcement, which could, in localised areas, overwhelm the CP system and thus cause a high level of corrosion. In this case, the solution could be to limit the length of the CP zone by having limited commoning of the segments. This would limit the stray current pickup path.

The DC negative circuit can be made in different configurations, such as a spur circuit, a ring circuit or a combination of the two. The designer should normally decide on the most economical arrangement by considering the size of the cables required and their number. Whatever arrangement is chosen, there should always be electrical redundancy. It is usually the case that the higher the requirement for reliability and life expectancy, the more DC negative connections to the steel reinforcement are used. These can be made in a number of ways such as thermite welding, pin brazing, electrical arc welding of a plate to the reinforcement, using a percussive nail gun, or drill and self-tapping screw. Each of these processes has its adherents, and the most important thing is to make sure that they are undertaken properly. Pin brazing is my favourite as it is fast, positive, easy to test (hit the stud with a hammer—if it does not fall off, it is OK) and reliable. Contractors are not normally so keen as they require the purchase of a pin brazer, which is a substantial investment. Whatever joint system is used, the DC negative connection should be covered by a non-conductive epoxy or mastic to prevent corrosion of the copper core. Normal good practice is to have at least two DC negatives per zone.

REFERENCES

1. Concrete Society Technical Report CSTR No 73. (2011). *Cathodic Protection of Reinforced Concrete*, the Concrete Society, Surrey, UK.
2. BS EN ISO 12696 (2012). *Cathodic Protection of Steel in Concrete*, British Standards Institute, London, UK.

Chapter 8

Design of a cathodic protection system for masonry

John Broomfield

CONTENTS

8.1 PROBLEM

Between the 1880s and the early to mid-twentieth century, many major buildings were built with steel frames and brick, masonry or terracotta cladding. Many of those that survive are in major land mark locations and are now on historic registers.

The designers and builders of these structures did not provide any significant corrosion protection to the steel frames. As a consequence, over the many decades since their construction, moisture has penetrated the external cladding leading to corrosion of the steel frame. This has led to cracking, displacement of bricks, stones and tiles and loss of section of the structural steelwork. Figure 8.1 shows corrosion damage to a large brick building due to expansive corrosion products generated on the steel frame.

Moisture ingress has occurred for a number of reasons including the porous nature of the cladding materials, ingress through mortar joints,

Figure 8.1 Vertical cracking caused by corrosion of the steel frame columns on a large, 1930s built brick building.

detailing of decorative feature that retain rather than shed water and lack of maintenance or maintainability of gutters and downpipes.

This problem has some similarities to corrosion of steel in concrete. It is discussed in detail in Chapter 1. There is a series of unique issues when considering applying impressed current cathodic protection (ICCP) to these structures, which are discussed in this chapter. One issue is that of working with historic (listed) structures that require conservation rather than repair.

8.2 REPAIR OPTIONS

Until the development of ICCP and suitable anodes for these types of structures, the only options were to try to exclude moisture and where damage was excessive to dismantle the cladding, repair, clean and apply protective coatings to the steelwork and then to reassemble the cladding with new materials where the old ones were beyond repair.

The extent of dismantling and the difficulty of reassembling stones or other cladding materials can be unacceptable to historic conservation officials. The use of ICCP, usually in combination with more conventional techniques, can lead to an approach that minimised the disturbance to the structure, which is the conservator's preferred approach.

It should also be noted that even if every effort is made to repoint all the joints between bricks and masonry, all drainage is renovated and all flashing and waterproofing repaired and improved, the nature of the building design and the materials may still lead to ongoing moisture ingress and corrosion.

ICCP can minimise the amount of intervention on a structure and provide long-term corrosion control, at least in the areas where there is corrosion-induced damage to the masonry. It is also reversible in that if a new technique was developed, the anodes and wires could be removed and minimal damage would have been done to the structure.

8.3 HISTORY OF APPLYING ICCP

The earliest recorded application of ICCP to a masonry structure was to the entrance colonnade of the Royal College of Science in Dublin in 1991. This was followed in the projects in London and Manchester in buildings and London underground stations. A number of major West End Department stores and other buildings have been protected, as well as an increasing number of projects in such U.S. cities as New York and Chicago.

8.4 DESIGN ISSUES FOR ICCP OF STEEL-FRAMED STRUCTURES

The anodes for ICCP are discussed in Chapter 5. The ones most useful for steel-framed buildings are the mixed metal oxide–coated titanium ribbon, which can be slotted into the mortar joints between stones, bricks or tiles, and probe anodes, which can be installed unobtrusively and their connecting wires run in the mortar joints.

8.4.1 Corrosion control versus facade damage limitation

The most import issue is that the impressed current cannot travel though air. It can travel through brick, most types of stone, terracotta, mortar and most other building materials that have some porosity. It cannot protect steel that is not encased or in contact with a material that can act as an electrolyte.

If the air gap is small and corrosion product fills the air gap, then current will flow at the point in time when the corrosion product starts to exert pressure on the brittle cladding material. Therefore, cathodic protection can be said to be effective in stopping damage to cladding due to corrosion of encased steelwork.

However, it will not stop the loss of section of the steelwork until any gaps between the steel and the cladding are sufficiently filled with corrosion product to allow the passage of the protective current. It is therefore essential to conduct a structural survey of the condition of the steel frame to ensure that any present or future loss of section of the structural steelwork in areas where current cannot reach the steel is acceptable or will be monitored to ensure that there is timely intervention if required.

There is an enormous range of materials used to infill behind cladding. These can range from rubble, bricks and poor-quality concrete to high-quality mortars that may be lime or Portland cement based on age. The electrolyte surrounding the steel frame must therefore be evaluated to determine the extent to which corrosion control can be achieved by ICCP on each structure.

8.4.2 Continuity of steelwork

Immediately after the issue of getting current to the steel to protect it, comes the issue of finding all the steel to make it continuous. This is fairly straightforward for most reinforced concrete structures where a cover meter should find the steel and continuity testing is straightforward. However, especially on earlier building, constructed around 1900, there appears to be more 'improvisation' on-site. Random steel clamps may have been inserted to hold stones in place. These may be difficult to identify as being present, hard to locate when they are present and also may present challenges in establishing an electrical connection once identified and located.

Major problems were found on the steel frame of a major London department store. Stones that cantilevered out over the street were secured by short steel beams connecting back to the main steel frame. The short steel beams were loosely bolted to the frame, presumable to make it easier for the masons to locate these stones. As a consequence, there was no guaranteed electrical continuity to all the steelwork. The contractor had to trial a number of options before identifying a method of making the connections with minimal damage and kept within programme and budget.

8.4.3 Cathodic protection zoning

The layout of zones is particularly tricky with steel-framed buildings. Some designers have large numbers of very small zones. Others use smaller numbers of larger zones or may have subzones that can be adjusted using potentiostats on anode strings. Anodes follow the steel frame and must be installed to ensure that adequate corrosion control is exerted to those parts of the steel frame that are exerting tensile stresses on the cladding materials. This may be the outward facing toe of the flanges of an 'I' beam or the flange face itself, depending on the orientation of the beam or column. Joints between columns and beams can be large and complex requiring extra anodes.

In some cases, detailed engineering drawings of the frame are available with each steel section drawn in detail. In other cases, these will have been lost or rendered illegible over the years and careful investigation along with reasonable assumptions will have to be made.

Figure 8.2 shows a simple ICCP system applied to the steel columns of a residential building in central London. Each column is a separate zone

Figure 8.2 Two zones of a simple impressed current cathodic protection system applied to a Georgian brick building in a conservation area in London.

with two reference electrodes installed per zone. The lighter mortar shows mixed metal oxide–coated titanium mesh ribbon anodes installed in the mortar joints after replacing damaged bricks. This system has four zones, one at each corner. The control system is a simple manually operated one.

8.5 LIME MORTARS

Although lime mortars were used in the original construction, it is important to use them in any rehabilitation work. This requires specialist stone masons. Lime mortars behave very differently from Portland cement–based mortars. They are more flexible and have different water transmission properties. Where structures with original lime mortars have been repointed with Portland cement mortar, there are often problems with moisture build-up behind the Portland cement.

However, ribbon and probe anodes need to be embedded in Portland cement–based mortar with high reserves of alkali due to the acidic reactions at the anode surface. Some compromise may therefore be necessary on the type of anode used and how it is embedded.

8.6 STAINING

A small number of projects have exhibited staining, particularly around ribbon anodes in ashlar joints between cladding stones. This appears to be due to the oxidation of iron. This may be due to iron in the matrix of the stone or iron from the tools used to remove the mortar before inserting the anodes.

It is therefore important to embed the anodes deep enough to minimise the risk of staining coming to the surface and to eliminate the use of steel-tipped tools for raking out mortar. Probe anodes do not seem to suffer from this problem as they are embedded deeper into the masonry. If the stone is known to have a significant iron content, then trials may be required.

8.7 WORKING ON HISTORIC (LISTED) BUILDINGS

With historic listed buildings, the approach is not one of repairing a structure and improving the aesthetics. A simple and non-exhaustive list of issues to be considered is as follows:

- All historic fabric are precious, not just the original, they help to tell the history of the structure, not just of its original construction
- Intervention should be avoided if at all possible and if it is necessary then it should be kept to a minimum
- Necessary intervention should be reversible if at all possible
- Any intervention (new materials) should be in addition to the existing structure and should not replace it
- Intervention should be carried out with sympathetic materials (good engineering)
- Avoid destroying historical evidence
- Make clear modern changes while being sympathetic
- Work closely with the listing authorities, planning authorities and owner to ensure that everyone understands others' requirements and priorities and that the solutions and compromises are agreed and recorded

8.8 TRIALS

The need to meet the requirements of historic listing authorities along with the issues raised in Section 8.7 means that small-scale trials are more likely to be required in historic steel-framed buildings to ensure that the system can be installed in a way that is visually acceptable and that is technically effective in passing current to the steel. Since reference electrodes must be

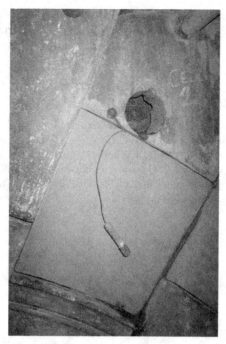

Figure 8.3 Damaged stones have been replaced before the cathodic protection contractor arriving on-site. A core hole is therefore required to make a connection to the steel and to install a reference electrode.

installed as well as anodes, wires must be run and connections made to all steelwork; it may be necessary to install connections and reference electrodes by coring through stones, making connections and replacing the cored piece of stone, as shown in Figure 8.3.

8.9 STANDARDS AND GUIDANCE

There are no current standards specifically on cathodic protection of steel-framed buildings and structures. However, the main standards such as BS EN ISO 12696 apply but with additional requirements such as those given in Chapter 2[1].

The best current guidance is the NACE International Publication 01210 *Cathodic protection of masonry buildings incorporating structural steel frames*, published in 2011[2]. This is a joint document with the Corrosion Prevention Association (CPA) in the United Kingdom and is available from the CPA website. This has a comprehensive list of examples and case histories as well as covering in detail the issues specific to this type of application of cathodic protection.

NACE has now started a new Task Group to Work on a Standard Practice for the inspection of early twentieth-century steel-framed, masonry-clad buildings before a standard on cathodic protection of these structures.

REFERENCES

1. BS EN ISO 12696 (2012). *Cathodic protection of steel in concrete*, British Standards Institute, London, UK.
2. NACE International (2011). *Cathodic protection for masonry buildings incorporating structural steel frames*, Publication 0210, NACE International, Houston, TX. Jointly published as Technical Note 20, Corrosion Prevention Association, Bordon, UK.

Chapter 9

Design of cathodic protection systems for new reinforced concrete structures

Richard Palmer

CONTENTS

9.1 INTRODUCTION

Reinforced concrete is an excellent material for cost-effective construction. The material is capable of achieving the required strengths and with care, the necessary durability. Nonetheless, it is clear today that reinforced concrete is vulnerable to damage from the environment. This is partly due to changes in the manufacture of Portland cement concrete. The indications are that these changes, aimed at developing higher strengths, have decreased the beneficial ageing effects found with older concrete. To reduce the risk of substantial deterioration within the structure's planned lifetime, designers are now including other durability enhancements such as cement replacements (fly ash, silica fume etc.), cathodic protection (CP) and coated reinforcement. The designer is free to incorporate one or more of these techniques according to the degree of deterioration risk considered acceptable. This chapter discusses the use of CP to enhance the durability of new reinforced concrete structures. Note that when applied to new structures, CP is designated as cathodic prevention by the current European Standard BS EN ISO 12696 [1], the principal difference being that a lower current density is required to protect reinforcement. For the purposes of this chapter, the abbreviation CP is used to designate both types of system, unless otherwise indicated.

Throughout the 1970s and 1980s, there was increasing awareness that environmental conditions would cause reinforced concrete to deteriorate, principally due to corrosion of reinforcement. The financial significance of this situation was demonstrated by the BRITE-EURAM research work [2]. This project, undertaken by leading research groups within the European Community, aimed to remove many uncertainties with regard to design for durability. As a result, today's design codes offer improved guidance with regard to design for durability.

Repair and maintenance are necessary considerations throughout the life of a concrete structure. Industry has responded with provision of these services at varying levels of sophistication. A large element of this work is aimed at remedies for reinforcement corrosion damage due to inadequacy of the concrete cover zone in protecting embedded reinforcement.

The principal cause of this corrosion damage is penetration of aggressive agents from the environment. Of these, the major problems arise from carbon dioxide in the atmosphere and chloride ions in de-icing salt or sea water.

Carbon dioxide penetrates pores in the concrete cover and combines with the pore water to produce carbonic acid. This action, termed 'carbonation', results in reducing the normal high alkali compounds found within fresh concrete to inert carbonates, penetrating the concrete cover zone as a carbonation front. Concrete so affected (or carbonated) no longer has the necessary alkalinity to maintain embedded reinforcement in a passive state, and hence, once the carbonation front has penetrated the steel reinforcement depth, corrosion can start, given the necessary supply of humidity and oxygen. This form of deterioration is relatively easy to evaluate and treat. Carbonated concrete is easy to identify by a simple on-site phenolphthalein test. Where the carbonation front has not reached reinforcing steel, anti-carbonation coatings can be applied to stop further attack. Spalling due to localised carbonation can be remedied by cutting out the damaged concrete to behind effected reinforcement and undertaking patch repairs in accordance with the guidelines given in BS EN 1504 [3]. The 'inert' nature of the carbonated concrete thus enables a generally straightforward repair approach. If carbonation penetration is extensive, an electrochemical treatment may be used, the choice of treatment dictated by the concrete characteristics. Re-alkalisation is best combined with application of an anti-carbonation coating to avoid further attack [4]. Alternatively, CP can be used.

Chloride ion attack of embedded steel reinforcement is far more difficult to combat. Chloride ions arrive at the concrete surface in solution, either in sea water or in de-icing salt, and are transported into the concrete pores by diffusion as described by Bamforth [5]. Chloride ions can penetrate even

well-designed concrete mixes and with time will build up around the reinforcing steel. At a critical concentration of chloride ions, corrosion will
commence. The penetration of chloride ions does not, as with carbonation,
result in a static and inert zone of damaged concrete. The chloride ions do
not create a static damaged zone but remain mobile and are able to catalyse
corrosion in sufficient concentrations. The variability of concrete properties, even within a single casting, and the high mobility of chloride ions
by diffusion within the cement pore structure enable chloride penetration
to variable depths and concentrations. It is difficult to identify chloride-
contaminated concrete on-site. On-site tests exist but are relatively costly
and less accurate than laboratory analyses. Furthermore, the continued
mobility of chloride ions makes it difficult to calculate the corrosion risk for
a given structure. The classical repair approach of removing and replacing
contaminated concrete from spalled areas has been found not to work due
to continuing chloride activity. Repair of reinforcement corrosion resulting
from chloride contamination therefore requires an electrochemical solution. A typical example is a CP installation to repair a major reinforced
concrete jetty facility for the loading and storage of freight containers in
Kowloon. This structure, designed in accordance with the codes of practice
of the day, was 16 years old when a major programme of refurbishment
was needed to repair extensive corrosion damage that threatened its stability. Patching of spalled concrete areas had already been tried but without
success. In 1993, a CP system was successfully applied to the structure,
without interruption to the clients' container business, and later extended
to give a total installation size of $22,500$ m^2. The system comprises a mixed
metal oxide (MMO) anode embedded within a cementitious overlay. The
remedial works have arrested the rapid corrosion decay of the reinforced
concrete substructure and provided an effective durability upgrade. The CP
system continues to function well after 19 years of service.

The subject of this chapter is the use of CP for enhancing the durability
of new structures.

The marine and de-icing salt environment can prove particularly harsh
for concrete structures, subjecting them to continuous cycles of salt water
wetting and drying. The vulnerable areas are those within the tidal and
splash zone where wetting cycles result in excessive build-up of chloride contamination within the concrete pore structure. High chloride concentrations
set up diffusion gradients allowing chloride ions to move into the concrete,
eventually arriving at the reinforcement surface. In sufficient quantities, the
chloride ions are then able to disrupt the normal passive conditions for steel
in concrete, causing reinforcement corrosion. The steel corrodes to occupy
a greater volume and exerts tensile stresses on the cover concrete resulting
in spalling. If allowed to proceed unchecked, corrosion damage can lead
to structural weakening and ultimately to catastrophic failure. The time
period from construction to initial corrosion damage will vary as a function

of type of cement, water–cement ratio, depth of cover and the local environment. To indicate the extent of the problem, periods of 10–15 years to first corrosion damage are typical for reinforced concrete in exposed locations. The time for major maintenance will depend on local environmental conditions and the particular structure. Continued operation of the structure is usually the major consideration, although aesthetics often influence the implementation of a maintenance programme.

When designing new structures, it is common to include additional durability enhancement in high-risk areas. An example of this is the tidal and splash zone of bridge piers in a marine environment, where the life cycle cost of additional protection is small in comparison to the cost of future access and repair. The technique of CP is ideally suited for this application as it can be designed to specifically target high-risk areas of the structure.

9.2 ALTERNATIVES TO CATHODIC PROTECTION

From the previous discussion, it is clear that the durability of reinforced concrete is largely a measure of the concrete's ability to control penetration of aggressive agents from the local environment. The principal area for improvement of concrete durability is hence the reinforcement cover zone. Concrete permeability is a fundamental characteristic for improvement as reductions in permeability considerably enhance the degree of protection. A further consideration is the ability of concrete to chemically neutralise and combine chlorides, thereby increasing the time taken for chlorides to initiate corrosion. The durability of reinforced concrete may be improved by these methods:

1. Modifying the concrete mix
2. Increasing the concrete cover
3. Enhancing the curing process
4. Applying physical surface barriers (coatings) to concrete
5. Applying electrochemical corrosion control systems or corrosion inhibitors
6. Applying reinforcement coatings or using corrosion-resistant reinforcement

The choice of durability enhancement should be selected on the basis of cost-effectiveness over the life of the structure, comparing the initial capital cost of treatment together with its effective lifetime and maintenance costs.

The first three methods are integral to design and production of the required concrete, and it is assumed that for a new structure the best practices will be employed. These include the use of cement replacements as described by Bamforth [6]. The use of replacements that can cause a large reduction in permeability should be undertaken with care as with this increase can come a reduction in ductility, which can, in a real structure, give an increase in the number and severity of cracks. This could allow chloride ions to get to the reinforcement very quickly. Concrete admixtures also fall within this group. There are a range of admixtures such as acrylic polymers, stearates and so on, which impart water reducing, waterproofing and other properties designed to improve various qualities of hardened concrete. The function of such admixtures is described in most textbooks on concrete such as that by Neville [7].

The impact of concrete curing on permeability is well documented [7] and it is clear that measures to ensure adequate curing are fundamental to optimising durability. A further technique utilising this principle is the use of controlled permeability formwork [8]. The ability of the formwork to optimise the quantity of water required for cement hydration results in a concrete cover zone of reduced permeability, together with an improved surface finish.

Coatings, the third treatment class, fall into two general categories. These are either penetrative or surface coatings. Penetrative coatings such as silane serve an essentially hydrophobic function, thereby resisting the uptake of chloride-bearing water. Surface coatings are available in many formulations. They comprise combinations of up to four constituents, a binder, inert fillers or pigments, liquid solvents/dispersants and additional additives for particular properties. Typical binders include epoxy and polyurethane resins, all of which have particular uses in the construction industry. The choice of coating constituents dictates the resultant performance characteristics such as adhesion, permeability, wear resistance, ease of application and cost. A large number of coatings are available, hence the engineer is recommended to review similar case studies and technical information for accurate lifetime and cost data. The life of a coating system typically varies from approximately 5–15 years depending on the material type and the environment. When comparing different systems, allowance should then be made for the cost of regaining access to the structure and of maintaining/renewing the coating. While coatings are advantageous for protecting and decorating buildings, they are less suitable for application to marine and other structures in humid environments. Furthermore, they do not address the mechanism of reinforcement corrosion due to chloride attack.

The last two classes of treatment, electrochemical systems reinforcement coatings and corrosion resistant reinforcement, adopt an alternative strategy. The contaminants are allowed to penetrate towards reinforcement but their corrosive effect within the concrete is either neutralised or reduced.

Electrochemical control systems fall into three categories; CP/prevention, chloride removal and re-alkalisation. Both re-alkalisation and chloride removal are technically similar to CP while using much higher applied current densities. They use temporary surface electrodes and specific electrolytes to restore alkalinity around reinforcing steel (re-alkalisation) or to move chloride ions away from the reinforcing steel and out of the concrete (chloride removal). Both techniques are designed for relatively short application periods and are ideally accompanied by subsequent coating application to avoid re-contamination.

The most widely used electrochemical technique is CP. The CP mechanism is more fully described in Chapter 4. Particular applications using MMO anode systems is reviewed in Chapter 5. Initial guidance on the use of electrochemical refurbishment techniques can be obtained from the concrete repair standard BS EN ISO 1504 [3] with a more detailed discussion given by the Corrosion Prevention Association Technical Note 2 [9].

Also within the category of electrochemical barriers are corrosion inhibitors. These work by producing electrochemical conditions at the rebar concrete interface that inhibit the development of corrosion cells. There are a number of inhibitors currently available. These comprise chemicals such as calcium nitrite, sodium monofluorophosphate and so on. A more detailed review of these systems and their relative performances is given elsewhere [10].

The last of the listed classes of treatment proposed for durability enhancement is to apply a barrier coat directly to the reinforcement. This process, referred to as Fusion Bonded Epoxy Coated Reinforcement (FBECR), entails the use of factory prepared and coated bar. The FBECR should be manufactured in accordance with the guidelines issued by the British Standards Institute [11]. It is important to use high-quality product and to be cognisant of the possible effects both of poor installation handling and of interaction with uncoated reinforcement. Other corrosion-resistant reinforcement solutions include stainless steel or stainless steel clad reinforcement [12]. Bronze reinforcement has also been trialled and used.

Much work has also been done developing non-metallic reinforcement. Several structures have been built using reinforcing bars made from fibre-reinforced polymer (FRP). This new composite will see increasing use as the architectural and engineering community become more confident of the long-term performance characteristics of FRP in concrete.

For a state-of-the-art review of future developments in this area, the reader is referred to the National Composites Network [13].

9.3 CATHODIC PROTECTION/PREVENTION

CP is a well-established technique for long-term protection of new and existing concrete structures exposed to corrosive conditions. A list of significant milestones in the history of CP is given in Table 9.1. The technology, which has been applied to various structures worldwide, enjoys a 30-year experience base. The use and specification of CP for reinforced concrete is described in the European Standard BS EN ISO 12696 [1].

Table 9.1 Milestones in cathodic protection applications to reinforced concrete

Date	Event
1824	Sir Humphrey Davy discovers CP after work on ship hull corrosion After a further delay CP was then used extensively for buried and submerged steel protection
1965	Development and evaluation of mixed metal oxide anodes (DSA)
1973	CP system applied to top deck of Sly Park Crossing Bridge Deck in California, United States. This coke/asphalt anode system operated for 11 years
1974	Ontario Ministry of Transportation installs coke/asphalt CP to bridge deck
1974–1975	California installs coke/asphalt systems to bridge decks
1975–1980	Federal Highways Authority establishes Demonstration Project installing bridge deck systems
1977–1984	Development and application of slotted CP systems for bridge decks
1981	Development and application of mounded grid conductive CP systems with cementitious overlay
1982	Development and evaluation of conductive coatings as anodes for concrete substructure CP in U.K. DoT
1984	Application of FEREX anode (copper cored carbon polymer)
1984	CP applied to first parking structure in the United States
1985	Application of mixed metal oxide anodes (DSA) to bridge decks and substructures worldwide
1985	Application of first U.K. conductive coating CP to substructure
1987	Conductive overlay applied to bridge deck in Virginia, United States
1987–1992	Application of CP to pre-cast post-tensioned bridge decks in Italy
2013	40th anniversary of concrete CP

CP is successful in treating attack by chloride because it operates upon the 'as found' contaminated state of reinforced concrete and subsequently modifies the electrochemical state causing corrosion so as to prevent further deterioration. CP thus avoids the expensive removal of large quantities of chloride-contaminated concrete and minimises the downtime associated with repair. It nonetheless provides a long-term rehabilitation method that requires only minimal maintenance.

Although there are many proprietary anode systems available today, this section describes the characteristics and uses of impressed current MMO anodes, specifically those using a titanium substrate. This anode type is alternately referred to as the DSA® anode due to its dimensional stability in operation. Discovered by Henry Beer, the anode was first patented in the United States and Europe between 1966 and 1973 [14,15]. These anodes were found to perform with great stability at very high applied current density levels in aggressive environments. Over the following 10 years, MMO anodes largely replaced graphite anodes for use in the chlor-alkali industry. The characteristics of this type of anode make it ideally suited for the design of longlasting impressed current anode systems, as required for cathodic prevention.

Those seeking a detailed description of MMO anode electrochemistry are recommended to refer the work by Trassati [16]. Briefly, the anode comprises a valve metal substrate (typically titanium) with a MMO coating electrochemically applied to the surface. A valve metal is one that will passivate (form a protective metal oxide and hence stop current flow) if connected in circuit as an anode. This property gives titanium exceptional corrosion resistance. Of course it also means that to persuade titanium to work as an anode, the surface has to be activated in some way such that current can flow. This is the role of the thin layer of MMO (electrocatalytic) coating applied to the titanium substrate. The coating consists of one or more oxides of the platinum group metals such as iridium, ruthenium, palladium and so on. It is applied to a prepared surface and heat treated to form a MMO film of exceptional electronic conductivity. It has the further benefit of being resistant to accidental current reversal and tolerant of AC ripple from power supplies. The combination of substrate metal and MMO coating provides a durable anode, physically tough, easy to handle, inert to corrosion attack and suited for long-term protection of reinforcement in concrete. The MMO anode surface is hard (around 6 on the Mohs scale) and hence resistant to abrasion. In this respect, the anode will withstand physical handling on-site including being subjected to shotcrete impact when the technique is used to encapsulate the anode.

The MMO anode initially found favour due to its ability to withstand high current densities in aggressive environments. In its original environment, electrolysis cells within the chlor-alkali process, the anode is required to operate for a period of 6 months to 1 year at an anode surface current density of up to 12 kA/m^2. The MMO anode coating is consumed at a very low rate during this process (hence the original designation of DSA or dimensionally stable anode) further indicating its suitability as an embedded anode for CP of reinforced concrete. When used in the context of cathodic prevention of new reinforced concrete, the anode should be designed to operate at a maximum surface current density of approximately 110 mA/m^2; at this current density, the anode manufacturers quote typical lifetimes of the order 40 years or more. The MMO anode will deliver a certain amount of charge (current × time) as determined by the applied coating and its operating environment. As the current density demand reduces during CP operation, the lifetime extends. MMO anodes are tested for suitability in accordance with NACE standard TM0294-94 [17]. This is an accelerated test designed to ensure that the anode will provide a minimum charge density of 38,500 Ah/m^2 (40 years at 108 mA/m^2) during its lifetime and will endure current reversal (fault) conditions for 1 month with no adverse effects, all within three different aqueous solutions to mimic various environments in concrete. The MMO anode exhibits a linear relationship between lifetime and current density as plotted on a log-log scale. Thus, for each anode type, it is possible to derive a relationship of the form:

$$\text{Log life (years)} = A - B \times \log \text{current density (A/m}^2)$$

where A is a constant and B is curve gradient.

Put simply, less current density from the anode results in longer lifetimes. Using the preceding characteristics, it is possible to design MMO anodes for long-life operation in reinforced concrete environments.

The lifetime versus current density relationship is important when designing CP systems for reinforced concrete. When current is applied, the following effects can be measured:

- An immediate shift of reinforcement potential to more negative values, thereby reducing corrosion activity
- A cathode reaction, which generates hydroxyl ions (OH$^-$) at the bar surface
- The migration of chloride ions away from the reinforcing bar surface

The latter two effects lead to a continual improvement in the environment around reinforcement with a corresponding reduction in the current density required for continued CP. Research [18] carried out at Imperial College, London, indicates that the generation of hydroxyl ions is the effect that plays the most significant role when applying CP to reinforced concrete.

Because of the improvement to the concrete environment generated by application of CP, the required anode output to protect existing contaminated structures reduces throughout the life of the installation. As the required output reduces, the lifetime of the anode is significantly extended. Given a lifetime of approximately 40 years for a MMO anode at maximum rated current density, it is clear that with reduction in required output the lifetime will become far greater. In the case of CP designs for new structures, the anode system is designed to protect against chloride reaching the reinforcement. In consequence, there is no great change in the required anode output with time and it is hence important to reduce the anode design output for the required life of the installation. With this provision, cathodic prevention systems can be designed with lifetimes compatible to those of the new structures.

9.4 DESIGN CONSIDERATIONS

MMO anodes are produced in various configurations by manufacturers based principally in the United States, Europe, the Middle East and the Far East. Typically, MMO anodes for CP of reinforced concrete are produced in an expanded metal mesh format. These are available in sheets of approximate width 1–1.2 m and also in a narrower expanded metal ribbon format at widths of between 10 and 25 mm.

The mesh anode format was initially developed for the CP of existing structures. However, it has subsequently been successfully used for protection of new structures. The anode is lightweight and easily handled. It is produced in various output grades by varying the ratio of actual anode surface area to the projected planar surface of anode. That is to say, if 1 m^2 of anode mesh has a measured surface area of 0.15 m^2 and the current density on the anode surface is limited to 108 mA/m^2 then the anode output is quoted as 108×0.15 = 16.2 mA/m^2 of concrete. Typically, mesh anodes are currently available in outputs of between 16 and 40 mA/m^2 at the quoted upper limit of applied surface current density. Mesh anode is applied in single or multiple layers to achieve the required current densities for protecting embedded reinforcement. An illustration showing the use of multi-layered anode material is given in Figure 9.1. In this photograph, a double layer of anode mesh can be seen fastened around steel piles during protection of an existing jetty structure at Kwai Chung in Kowloon, Hong Kong. Note that it is also possible to combine mesh and ribbon anode materials. Electrical current is generally introduced to the anode mesh through a primary gridwork of current distributor (CD) bar. This material is fabricated from grade 1 titanium and supplied in strips of between 10 and 15 mm width. It is joined to the anode mesh by spot-welding using a standard electrical resistance spot-welder.

While the use of multi-layer anodes provides a means of varying the current density to correspond to differing reinforcement densities, research has shown that the application of n layers does not provide a full $n \times$ increase over the single layer output [19]. Investigation and practical experience indicate that the introduction of subsequent anode layers can reduce the maximum layer efficiency to between 60% and 80%.

Figure 9.1 Kwai Chung jetty.

A further important consideration applies to achieving a uniform current density at the reinforcement within each anode zone; this relates to the CP circuit resistance. The preceding example of varying current density by super-imposing anode layers, in other words increasing the active anode surface, only applies when the anode to reinforcement spacing is low (of the order 2–5 cm) and the anode overlay mortar has a low resistivity value (15–30 kΩ-cm). With geometry such as this, the anode/mortar resistance will represent a significant percentage of the total anode to steel-reinforcement resistance, hence changes in its surface will have an effect on the current density seen by the steel. When the anode to reinforcement separation is greater (5–25 cm or more), design of the anode system needs greater care. The result of this greater separation, coupled with a high resistance concrete mix, will change the current density dynamics considerably. The anode/mortar resistance will become small with regard to the total such that changes in anode surface will not affect the current density seen by the steel. When designing for this latter scenario, the introduction of balancing resistors to the CP circuit will be necessary to actively achieve the optimum current density at the steel surface.

The anode may be fixed to a prepared concrete surface and then secured by encapsulation within a cementitious overlay. Alternatively, it may be cast into the concrete element. The recommended external cover to the MMO anode is not less than 10 mm. It is important to note, however, that in locations where the anode will be embedded in an offshore environment such that it will some-times be below the surrounding water level, additional precautions should be taken. In this situation, it is necessary to avoid the possibility of CP current flowing out through low-resistivity sea water to the reinforcement rather than passing directly through the concrete electrolyte. Insufficient amounts of con-crete cover to the anode will allow the current to take the lower resistance return path, hence creating a condition of local high current flow from the anode. This will reduce both the current density provision to surrounding reinforcement and the anode lifetime. In this situation, it is necessary to insu-late the anode from the lower resistance sea water. This can be achieved by specifying a thicker layer of high resistance overlay. In a 1991 report describ-ing CP installation to the Tay Bridge [20], it was concluded that concrete cover to the anode be increased to greater than 20 mm and that a well-cured and high resistivity concrete be used. An illustration of this successful project is given in Figure 9.2. This shows the base of a bridge pier, which is being strengthened and provided with CP. The anode is fastened deeply within the structure before pouring concrete. The anode and additional reinforcement were cast in concrete in one operation. A subsequent review of CP applied to this structure by Glass [21] reports that in 1996, CP is providing continued protection while demonstrating that the technique is improving the environ-ment for existing reinforcement. Various solutions have been used to insulate the anode. These have included the use of cast in place fibreglass or glass fibre cement shuttering panels. The solution will depend on the particular project.

Figure 9.2 Tay Bridge pier.

For CP of new (or old) structures, the anode is fastened with the aid of special non-metallic fixings before encapsulation or casting-in by concrete. It is essential to avoid electrical contact between anode and cathode (reinforcing steel). A minimum spacing of 13 mm is recommended between anode and reinforcement. Particular details for installation are given in the model specification published by the Concrete Society [22].

Concrete separating the anode and reinforcement (cathode) provides the electrolyte that allows current to flow through the CP circuit. It is required to provide a secure means of fastening the anode before and after encapsulation and also to form an uninterrupted path for ionic current to pass between it and the reinforcement.

The preceding discussion indicates nonetheless that careful consideration of CP circuit resistance is needed when designing the anode system to achieve the desired uniform current density and hence ensure that all targeted reinforcement is protected.

The ribbon anode type is well suited to the CP of new concrete structures. The anode may be fastened to the reinforcement cage during assembly using special non-conducting fastenings. For reasons of economy and uniform current distribution, the general practice is to connect the anode to a primary current distribution network of current distribution (CD) bar. The ribbon anode is spot-welded to the CD bar. The anode spacing is calculated on the basis of required current density at the rebar. One should ensure that the anode ribbons are not too widely spaced. Using higher output ribbon at too great a spacing leads to an unacceptably uneven current distribution across the reinforcement cage. Applied research indicates that a maximum ribbon spacing of between 200 and 400 mm centre to centre should be adopted depending on parameters of the structure to be treated [23,24].

Ribbon anode is easily and rigidly fixed to the reinforcement cage so as to present a minimum profile to flowing concrete. Thus, the possibility of anode displacement during the concrete pour is minimised. Nonetheless, it is recommended that the anode is electrically monitored during the concrete pour to ensure that no short circuits develop between it and reinforcement (the cathode). This straightforward operation is carried out with a high impedance voltmeter. Simply connect the voltmeter between anode and cathode in the area of the pour and set initially to the 0–200-mV measurement range. If the anode installation has been done correctly, there will be no contact between the anode and cathode and hence no initial voltage reading. Before and during pouring concrete, it is customary to place a wet sponge between anode and reinforcement. This will enable the meter to register potential difference across the cell and so created between the anode and reinforcement. As soon as the concrete pour starts, there will be a steady change in recorded potential difference as the concrete electrolyte encloses more of the anode and reinforcement surface area. A short circuit condition is marked by a dramatic switch to a potential difference of or very close to 0 mV. Experience in recognising this situation is easily gained. In the event of a short circuit condition, it is necessary to temporarily halt the concrete pour and displace the anode local to the pour position until the short circuit condition is removed.

In the event of a short circuit condition being discovered after the concrete has hardened, a number of solutions are available. The first approach is to apply a high current density, approximately 2 A/m² of concrete for a period of not greater than 1 minute. This action will generally destroy the short circuit contact. In the event of continuing short circuit conditions, it is necessary to locate the fault. This may be achieved by measuring and plotting the anode potential contours across the surface of the CP zone containing the short circuit. The potential contours indicate the location of the short circuit. Once located, the area can be broken out, the fault repaired, and the concrete reinstated.

Typically, the current density provided at the MMO anode surface is limited to a value of 10 mA/ft² of anode (108 mA/m²) resulting from early work by the Federal Highways Authority in the United States. This current limitation exists to avoid any risk of damage to cement paste in contact with the anode surface as a result of acid generation. While this value is generally adhered to, it should be noted that the MMO anode manufacturers have commissioned independent test programmes that verify that theses anodes can be run at higher surface current densities, of the order 400 mA/m², for a limited time without damage to the anode or surrounding concrete.

The preceding consideration does not apply when designing anode systems for new structures; the initial current demands are uniform (new concrete with no chloride) and hence there will no inherent reduction in anode output. The requirements when designing an anode system for a new structure are as follows:

- Selection of anode zone based generally on the local environment and on ease of control of the CP system
- Calculation of the steel reinforcement area influenced by the CP
- Calculation of the required current density for CP
- Design of the anode system to ensure uniformity of current density applied to reinforcement for the required lifetime

The guidance given by the Concrete Society [22] is for anode zone sizes to be limited to between 200 and 500 m² of concrete in size. It is not possible to be precise with this figure as it depends on the particular structure. Say, for example, a pier is to receive CP and that there exists a small zone around the pier footing where the concrete is wetter due to capillary rise from groundwater, then that area will generally require a higher current density for protection. In this instance, it is preferable to create a separate CP zone at the base to allow better control of the installation. Selection of anode zones is largely a question of experience, taking into consideration the structure and the local environment.

The maximum anode to reinforcement spacing for effective CP has been evaluated by several researchers. The majority of research in this area has been directed at the penetration of current in existing concrete structures. Hunkeler [25] presents a resistivity model for use in calculating the division of current density between top and bottom rebar in the case of a reinforced concrete slab, following his work evaluating a CP installation at the San Bernardino Tunnel in Switzerland. His site measurements indicate that for an approximately 300-mm-thick slab, approximately 70% of the current will flow to the top mat of steel with the remaining 30% to the bottom. He also presents a model for calculation of the current penetration, which again uses the concrete resistivity as the controlling factor. This is in general agreement with the work of Bennett [26] and Pedeferri [27].

Bennett's work, also directed at existing concrete structures, shows the current density required for CP to be proportional to the level of the chloride contamination. Using Bennett's example, the CP current for the zone is calculated using the formula:

$$\text{mA/m}^2_{\text{concrete}} = (\text{mA/m}^2_{\text{top mat steel}}) \times (\text{m}^2_{\text{top mat steel}}/\text{m}^2_{\text{concrete}})$$

for double mat of steel divide mA/m²$_{\text{concrete}}$ by 0.7 to get total current requirement per concrete area

However, Pedeferri's work [27] relates specifically to new structure CP. He presents data that clearly indicates that CP currents will flow to far greater depths in new concrete structures due to the absence of chloride and the consequent corrosion activity. Italy has been foremost in the development of this application having installed approximately 100,000 m^2 of CP to several new post-tensioned highway viaducts in Italy [24]. The PIARC Technical Committee on Road Bridges [28] published a document in 1991, which acknowledges this technique for new structures and gives broad guidelines for its use. To summarise, current literature indicates that reinforcement within non-chloride-contaminated concrete, that is, new structures, and located at depths of up to 40 cm from the anode may be cathodically protected. Pediferri's work indicates that reinforcement at greater distance from the anode may be protected.

The current requirement per area of reinforcing steel may be graphically estimated from knowledge of the chloride concentration. Later work underway at Imperial College [18] confirms this general relationship while further demonstrating the significance of the electrochemical displacement of chloride ions and generation of alkali conditions, further important effects of applying CP to reinforcement in concrete.

Practical evidence of the relationship between chloride level and the current density required for is reported for an approximately 100,000-m^2 CP installation to several new post-tensioned highway viaducts in Italy [24]. Data showed that the polarisation criteria is met in the case of uncontaminated new concrete structures by current densities between 1 and 2 mA/m^2 steel with concurrent voltage requirements of 2–3.5 V. Conversely, in the case of high chloride levels (1–3% by weight of cement), current densities of between 15 and 20 mA/m^2 of steel area are required with typical corresponding voltages of the order 10 V.

Design considerations such as cable connection integrity, reference probe selection and so on are common to all CP systems and are dealt with in Chapter 7.

9.5 EXAMPLES OF NEW STRUCTURE CP

While some project references have been made in the preceding text, the following section provides further examples of particular CP installations to reinforced concrete of new structures or newly cast elements.

The aforementioned Tay Bridge has been undergoing a programme of CP repairs for which evaluation work started in 1986. While this is clearly an existing structure, the CP installations have entailed installing a CP system comprising a MMO mesh anode fixed to a prepared concrete substrate beneath supplementary reinforcing steel. Refer back to Figure 9.2 for installation. Both the anode and additional reinforcement were then cast within fresh concrete. Using this technique, the base of bridge piers within the tidal and splash zone has been increased in section while at the same time a CP system has been introduced to enhance the durability. Another U.K. installation of CP to new reinforced concrete was the Felixstowe Ro-Ro ferry bridge project carried out in 1991. For this structure, a combination of mesh and ribbon anode was cast into the reinforced concrete to provide continuous corrosion protection.

In France, an innovative CP installation was carried out in 1989 to provide CP to a new 650-m² bridge deck at Hauteville-sur-Fier. A MMO mesh anode was embedded during prefabrication of deck slab units. This was easily achieved during the pre-casting process. After pouring concrete to adequately cover the top reinforcement layer, panels of mesh anode together with pre-welded current distributor bar were placed. Figure 9.3 shows the mesh anode being placed and clamped to the pre-casting frame before the final concrete pour to level. The pre-cast slabs were then installed onto a steel sub-structure. After placing the slabs the anode panels were electrically connected on-site prior to finishing the deck. The system was subsequently energised to provide corrosion protection.

Italy has made extensive use of CP for preventative maintenance; a significant number of new motorway bridges incorporate cathodic prevention of the decks and parapets. The majority of these new bridges are along the A32 Turin–Frejus motorway. The bridges are all constructed from pre-cast post-tensioned reinforced concrete box girders. CP has been applied in a variety of different and innovative ways: either to box girder units following casting or to completed bridge structures. The optimised

Figure 9.3 CP anode integrated within precast deck slab.

CP application to box girder units took place as a separate operation in the pre-casting facility. Immediately after casting, a MMO mesh anode was bonded to the deck surface by encapsulation within a polymer modified overlay. Figure 9.4 shows the box girder units in the pre-casting shed complete with the applied mesh anode, immediately before overlay application. Once the overlay had been placed, the units were transported to site and launched. Following completion of civil engineering works to erect the span, the individual CP units were electrically connected. The reinforced concrete bridge parapets were also provided with CP. The optimum method of installation to these units was to secure a MMO ribbon anode to the reinforcing cage and then cast concrete using movable formwork. This installation is illustrated by Figure 9.5, which shows a section of reinforcement cage complete with attached anode in front of a recently cast cathodically protected parapet. By this method, minimal preparation work, high output and a good quality finish were assured. The electronics for both power supply and microprocessor control of these CP installations are located within the box girders. Specific computer software provides safe control of the installation with facilities for remote monitoring.

In the United Arab Emirates, there are two notable CP installations to new structures; the first being CP of the structural reinforced concrete frame of the Juma Bin Usayan Al-Mansouri building, Abu Dhabi, and the second being CP of the replacement coping to the quayside of Port Rashid in Dubai.

In the United States, a number of American bridge decks have been rebuilt to incorporate CP, for example the 1300-m^2 CP installation to the

Figure 9.4 Italian bridge segments with integrated deck CP.

Figure 9.5 CP integration to new bridge safety barriers.

deck of the Old Lyme Bridge in Connecticut. This structure was protected using ribbon anode.

More recently, a CP system has been cast into the piers and columns supporting the Rambler Channel Bridge in Hong Kong. This structure carries the MTRC rail link to the new Lantau Airport facility. The client requested the inclusion of CP to enhance the durability of high-risk zones of this key communication link, namely the tidal and splash zone of reinforced concrete piers in sea water.

The practice of including CP to enhance durability of key reinforced concrete elements has even extended to recent refurbishment work to the Sydney Opera House substructure. This well-known Sydney landmark needed maintenance work to the reinforced concrete structures supporting the visitor walkways that encircle the Opera House. The foundation and substructure elements of the Opera House were constructed in the mid-1960s and have deteriorated over time due to chloride penetration into the reinforcement. The refurbishment work on the Opera House includes CP to new and existing concrete elements.

Today, the biggest user of cathodic prevention is Saudi Arabia, which is applying ribbon anodes to several hundred thousand square metres per year.

Table 9.2 lists a number of new reinforced concrete structures where CP has been used as a means of increasing durability.

Table 9.2 New reinforced concrete structures or elements with cathodic protection installed

Project	Location	Country	Type of Structure	Anode Type	Area (m²)	Date Installed
Mt. Newman Mining – Phase I	Port Hedland	Australia	Jetty	Mesh	2,100	1989–1992
Sydney Opera House	Sydney	Australia	Beams, columns, North & West Broadwalk	Internal	150	1996
Swimming Hall	Holbeak	Denmark	Swimming Pool, Edges[a]	Ribbon	290	1990
Swimming Hall	Svedenborg	Denmark	Swimming Pool, Edges[a]	Ribbon	30	1990
Hauteville sur Fier, Bridge	Hauteville sur Fier	France	Bridge Deck Units (Prefabricated), Curbs	Mesh	630	1989
Tung Chung Bridge	Kowloon	Hong Kong	Pile Caps and Columns	Ribbon	1,200	1994
Ranbler Channel Bridge	Lantau	Hong Kong	Piers and Columns	Ribbon	7,375	1996
Autostrada Torino-Frejus, Viaduct	Brunetta	Italy	Post-Tensionned Bridge Deck, Curbs	Mesh	12,080	1991
Autostrada Torino-Frejus, Viaduct	Brunetta	Italy	Curbs	Ribbon	4,330	1991
Autostrada Torino-Frejus, Viaduct	Clarea	Italy	Post-Tensionned Bridge Deck	Mesh	15,200	1989
Autostrada Torino-Frejus, Viaduct	Deveys	Italy	Post-tensionned Bridge Deck	Mesh	4,830	1991
Autostrada Torino-Frejus, Viaduct	Deveys	Italy	Curbs	Ribbon	1,730	1991
Autostrada Torino-Frejus, Viaduct	Giaglione	Italy	Post-Tensionned Bridge Deck, Curbs	Mesh	13,860	1988
Autostrada Torino-Frejus, Viaduct	Passaggeri	Italy	Post-Tensionned Bridge Deck	Mesh	8,030	1992
Autostrada Torino-Frejus, Viaduct	Pietrastretta	Italy	Post-Tensionned Bridge Deck	Mesh	4,830	1992
Autostrada Torino-Frejus, Viaduct	Pietrastretta	Italy	Curbs	Ribbon	1,730	1991
Autostrada Torino-Frejus, Viaduct	Ramat	Italy	Post-Tensionned Bridge Deck	Mesh	15,660	1991
Autostrada Torino-Frejus, Viaduct	Ramat	Italy	Curbs	Ribbon	5,620	1991
Autostrada Firenze-Bologna, Bridge	Rioveggio	Italy	Bridge Deck, South Bridge[a]	Mesh	1,580	1987

(Continued)

Table 9.2 (Continued)   New reinforced concrete structures or elements with cathodic protection installed

Project	Location	Country	Type of Structure	Anode Type	Area (m²)	Date Installed
Autostrada l'Aquila-Grand Sasso, Viaduct	S.Nicola 2	Italy	Post-Tensionned Bridge Deck, Curbs	Mesh	2,800	1990
Autostrada l'Aquila-Grand Sasso, Viaduct	Situra	Italy	Post-Tensionned Bridge Deck, Curbs	Mesh	2,200	1990
Autostrada Torino-Frejus, Viaduct	Venaus	Italy	Post-Tensionned Bridge Deck, Curbs	Mesh	13,130	1991
Sumitomo Mining Industry	Ehime	Japan	Beam (New PC Beam)		45	1990
Port Rashid	Dubai	U.A.E.	Crane Rail Support Slab[a]	Mesh	465	1994
Juma Bin Usayan Al-Mansouri, Building	Abu Dhabi	U.A.E	Walls, Rib Slabs, Stairs, Columns, Beams	Mesh	3,720	1991
Felixstowe Ro/Ro, Bridge No. 3	Portsmouth	UK	Deck	Mesh	400	1991
Felixstowe Ro/Ro, Bridge No. 3	Portsmouth	UK	Beams	Ribbon	100	1991
Tay Bridge	Dundee	UK	Bridge Columns[a]	Mesh	120	1987
CTDOT – Rte I-95 over the Lieutenant River	Old Lyme	USA	Bridge Deck – Ribbon	Ribbon	1,301	1990
Mid-Hudson Bridge	Poughkeepsie	USA	Bridge Deck[a]	Mesh	560	1989
World Trade Centre	New York City	USA	Parking Garage	Ribbon	186	1993
World Trade Centre	New York City	USA	Parking Garage	Ribbon	604	1994

[a] New element on old structure.

9.6 OPERATION AND MAINTENANCE

Once the CP system has been installed, it is necessary to operate and maintain it. Operation of the CP system involves an element of routine visual inspection. This is simply done as a general maintenance operation involving simple checks to ensure no physical damage to the installation. Maintenance of the electrochemical function is the work of specialist CP personnel. This involves periodic checks of system response in accordance with the relevant standards.

To make this task easier, many of the electrochemical control operations can be automated and operated by specific software, for example, the operations to periodically monitor reinforcement potentials and reset applied current densities against given criteria. Not only can the reporting functions be automated but a number of error conditions, such as interrupted, or excessive current flow, can be signalled.

The electronics used to operate and control CP systems have borrowed many functions from automatic process plant control as well as from advanced network theory. This latter addition allows large CP installations to be fully remotely controlled either by trained on-site maintenance staff or by corrosion engineers in another town or even another country.

CP systems can be designed with acceptably long lifetimes. Anode lifetimes of 40 years or more are easily attainable. The associated hardware, that is, the cables, electrical connections, reference electrodes and electronic equipment for power supply and monitoring, are more vulnerable to breakdown, but with careful design, these components can be sufficiently durable and are easily replaced where the design allows.

The electronic equipment used to power and monitor a CP installation can be designed for lifetimes of approximately 20 years. Equipment should be chosen for ease of maintenance. Some manufacturers use modular designs, which readily allow replacement of circuit boards in the event of premature failure.

9.7 ECONOMICS

An impressed current CP system comprises the following:

- An anode
- A power supply
- A monitoring system

The specifier has several choices for each of the preceding point, depending on the required durability, the level of maintenance proposed and finally the available budget. For inclusion into new structures, it is necessary to allow for the cost of fixing the anode, cabling and other hardware. In addition, it is important to have a certified corrosion technician in attendance during certain operations, such as checking electrical connections, pouring concrete and system commissioning.

The approximate 1995 materials cost of a typical CP system designed to protect around 5000 m² of jetty substructure was as follows:

- CP anode: US$ 35/m²
- CP system: US$ 70/m²

The CP system comprises the anode and all electrical and electronic components.

These are values based on 1995 tender prices for large projects. The CP system cost comprises typically 10% of the overall project price.

Cathodic prevention costs today are somewhat lower. A typical materials figure (with no installation costs) for a large Middle Eastern system in 2011 for the foundation slabs of a power and desalination plant of 150,000 m² with a MMO ribbon and an automatic remote power supplies was as follows:

- −CP anode: US$ 15/m²
- CP system: US$ 50/m²

While reductions of scale apply to the material costs, one notes that the overall system cost does not reduce by the same percentage. The system cost includes the electronic power supply and control units. These components are both less numerous per system and, today, tend to incorporate more sophisticated electronics to allow networked remote control.

The technique of CP provides a secure method for long-term corrosion protection to vulnerable areas of new reinforced concrete structures.

REFERENCES

1. BS EN ISO 12696. '*Cathodic Protection of Steel in Concrete*', International Standards Organisation, Geneva, Switzerland, February 2012.
2. BRITE-EURAM Project, 'DURACRETE', Duracrete Final Technical Report. Brite Euram Project, Gouda, CUR, May 2000.
3. BS EN 1504, section 9. '*Products and Systems for the Repair and Protection of Concrete Structures*', International Standards Organisation, Geneva, Switzerland, July 2009.
4. CEN/TS 14038-Part 1. '*Electrochemical Re-alkalization and Part 2 Chloride Extraction Treatments for Reinforced Concrete*', International Standards Organisation, Geneva, Switzerland, 2004.

5. P. Bamforth. 'Admitting that Chlorides Are Admitted', *Concrete*, November/December, 1994.

6. P. Bamforth. '*Specification and Design of Concrete for Protection of Reinforcement in Chloride Contaminated Environments*' UK Corrosion and Eurocorr 94, Bournemouth, UK, 1994.

7. A.M. Neville. '*Properties of Concrete*' 5th edition. Pitman, London, UK, 2011.

8. Price W.F. The improvement of concrete durability using controlled permeability formwork. Proceedings 5th International Conference, Structural Faults and Repairs, Vol. 2, 1993, pp. 233–238.

9. J. Drewett, J. Broomfield. '*An Introduction to Electrochemical Rehabilitation Techniques – Tech Note No 2*' CPA, Hampshire, UK.

10. S. Dressman, T. Osiroff, J.G. Dillard, J.O. Glanville, R.E. Weyers. '*A Screening Test for Rebar Corrosion Inhibitors,*' Transportation Research Board, Paper No 91, Washington, DC. January 1991.

11. BS. 7295 'Epoxy Coated Rebar'.

12. S. Yeomans. '*Galvanized Steel Reinforcement in Concrete*' 1st edition. Elsevier Science, November 2004.

13. The National Composites Network, http://www.ncn-uk.co.uk/.

14. H.B. Beer, US Appl 549 194, 1966.

15. O. De Nora, A. Nidola, G. Trisoglio, G. Bianchi, Brit Pat 1 399 675, 1973.

16. S. Trassati. '*Electrodes of Conductive Metallic Oxides*' Elsevier, Amsterdam, the Netherlands. 1981.

17. NACE. '*Testing of Embeddable Anodes for Use in Cathodic Protection of Atmospherically Exposed Steel-Reinforced Concrete*' NACE Standard TM0294-94. NACE, Houston, TX, 1994.

18. G.K. Glass, N.R. Buenfeld. 'On the Current Density Required to Protect Steel in Atmospherically Exposed Concrete Structures', *Corrosion Science* Vol 37, No 10, pp. 1643–1646, 1995.

19. T. Pastore, P. Pedeferri, L. Bolzoni. '*Current Distribution Problems in the Cathodic Protection of Reinforced Concrete Structures*' Dipartimento di Chimica Fisica Applicata, Politecnico di Milano, Rilem Conference, Melbourne, Australia. August/September 1992.

20. A. Watters. '*Cathodic Protection of Tay Road Bridge Substructure*' Construction Maintenance and Repair Magazine, January/February, 1991.

21. G.K. Glass. 'An analysis of monitoring data on a reinforced concrete protection system', *Materials Performance* Vol 35, No 2, pp. 36–41, 1996.

22. '*Model Specification for Cathodic Protection of Reinforced Concrete*' Concrete Society and Corrosion Engineering Association Report No. 37. The Concrete Society, London, UK.

23. '*Cathodic Protection of Reinforced Concrete Bridge Elements: A State of the Art Report*' SHRP-S-337. The Strategic Highway Research Program, Washington DC. 1993.

24. M.A. Biagiolli, M. Tettamanti, A. Rossini, L. Cassar, G. Tognon, G. Familiari. '*Anodic System for Cathodic Protection of New Reinforced Concrete Structures: Laboratory Experience*' NACE Corrosion Conference 1993. Paper 321.

25. F. Hunkeler. 'Etudes de la protection cathodique du beton armé dans le tunnel San Bernardino' Octobre 1992, Département fédéral des transports, des communications et de l'energie, Office fédéral des routes, Switzerland, 1992.

26. J. Bennett, T. Turk, '*Criteria for the Cathodic Protection of Reinforced Concrete Bridge Elements*' SHRP-S-359. The Strategic Highway Research Program, Washington DC. 1994.
27. P. Pedeferri. '*Cathodic Protection of New Concrete Constructions*' International Conference, 2–3 June 1992. Institute of Corrosion in association with IBC Technical Services Ltd.
28. P. Baldo. '*Cathodic Protection of Bridge Decks*' PIARC Technical Committee on Road Bridges. 1991.

Chapter 10

Power supplies

Paul M. Chess and Frits Gronvold

CONTENTS

10.1 INTRODUCTION

When cathodically protecting steel-reinforced concrete structures, the need for power is normally modest compared to that required for cathodically protecting steel structures in water or soil. For concrete, it is normal to use several small power supplies, each with an output of around 0.5–5 A and a maximum output voltage in the region of 8–10 V depending on the anode type. This can be compared to traditional cathodic protection (CP) systems where 200 A transformer rectifiers are common and single power supplies can have a capacity of 1000 A or more at up to 50 V. When protecting masonry structures, the difference is even greater as the current requirements are very low and the output is often in the region of 10–500 mA at around 4 V.

The types of power supplies that are relevant for CP of steel in concrete generally require a higher degree of control in their operation than that in 'traditional' CP. For these reasons, the traditional tap-changing switched transformer rectifier, variac-controlled and thyristor power supplies are not normally used. The 'typical' traditional manual system uses thyristor-controlled transformer rectifiers as these are fairly efficient and can have some interface with electronics. With the advent of the microprocessor, there has been a general move to different technologies such as linear and switched-mode power supplies. The reasons are space efficiency, cleaner output, simpler to interface with other electronics and lower electromagnetic emissions. This change in power supply technology has been accelerated for reinforced concrete CP systems by the increasing popularity of remote and computer-controlled installations. These driving forces have produced significant differences in the physical construction of 'a traditional' power source and a 'reinforced concrete' power source.

The objective of this chapter is to outline the principles behind each of the popular power supplies, explain the choices that a CP designer should be aware of when specifying power supplies and, finally, illustrate the design of a typical remotely controlled power supply system for a CP project.

10.2 TYPES OF POWER SUPPLIES

For CP of steel in concrete or masonry, a relatively small direct current (DC) delivered between 1 and 10 V is normally required. Generally, this is obtained by transforming and rectifying a mains electricity supply; but power may also be delivered from batteries charged by solar cells, windmills and other electricity generators. The reinforced concrete structures where CP is to be made normally have some form of mains electricity available, and only this will be further considered.

There are two forms of power sources available, that is, single phase and three phase. In general, concrete power supplies tend to use a single-phase supply due to their limited current output requirement. Sometimes, when there are long alternate current (AC) cable runs the cabinets are fed consecutively by single phase on a three-phase cable. In public structures such as swimming pools, where power supplies and their feed wires may be near the public, it is not uncommon for the CP designer to require that a step down transformer (which is normally 24 or 48 V) be used to supply the various localised power supply units (substations) that are distributed around the structure.

All the various types of power supplies are designed to reduce the voltage and convert the AC into a DC output. The CP engineer may require that the supply can adjusted in current output level, or in voltage output level with a limiter on the output. Sometimes, more interactive forms of adjustment are required such as potentiostatic or potential decays. The various systems that are commercially available as power supplies are discussed in the following subsections. The correct power supply type for a particular application should be considered by the CP engineer. The parameters that should be considered are reliability, control, efficiency, size and output ripple.

10.2.1 Manual tap transformer + rectifier

A power supply comprising a transformer + rectifier is a simple and very robust unit. A transformer inputs the AC mains voltage and reduces the voltage to a desired amount depending on the proportion of windings around a soft iron core. Adjustment in output voltage level is obtained by using a switch to choose between different outlets, that is, the various separated windings, of the transformer. These are called 'taps'. Finer adjustment can be obtained by using a moving coil transformer where the direct mechanical switches are replaced.

The reduced-voltage AC from the transformer is then passed through a 'bridge'-type circuit where the current is rectified, that is, converted into a DC output (Figure 10.1).

It should be noted that the output power from a unit comprising a transformer and rectifier is not a pure DC but only a rectified AC, which may feel like electric shocks even at very low voltages. Some anode materials are reputed to be damaged if there is a lot of ripple in the current (ripple is the amount of change in waveform). This is the primary reason why the amount of ripple allowed is specified in a standard CP power supply specification. The secondary reason is that it makes interference on the circuit very difficult to track down. These units have disappeared from the European market but are still commonly used in the United States.

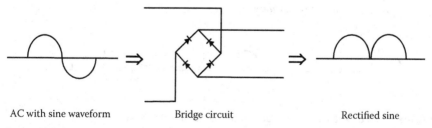

AC with sine waveform   Bridge circuit   Rectified sine

Figure 10.1 Full-wave rectification using a bridge circuit.

10.2.2 Transformer + rectifier + smoothing circuit

The AC supply circuit is passed into a transformer, which reduces the voltage. Various output voltages are obtained by having separate windings on the transformer and by energising or removing these from the circuit. The low-voltage alternating current is then passed through a bridge circuit so that it is rectified. This is then smoothed using electrolytic capacitors as shown in Figure 10.2.

The current can be smoothed out with a single capacitor, as in circuit A in Figure 10.2, or by several capacitors in so-called LC links, as shown in circuit B in Figure 10.2.

Normally, electrolytic capacitors are used as they are cheap and have a high capacity. The service life of electrolytic capacitors is relatively short, and they are normally the life-determining part of this type of power supply. Their popularity has also waned to the extent that it is difficult to find manufacturers for concrete-optimised units.

10.2.3 Thyristor Control

A thyristor can be made to act as a controlled rectifier. The output level is adjusted by placing these electronic devices in the rectification circuit and controlling the conduction, which the device either prevents or allows. This is shown in Figure 10.3 for full and partial conduction.

The amount of current passed depends on when the phase angle control unit of the thyristor is energised. This in turn is energised by a DC input;

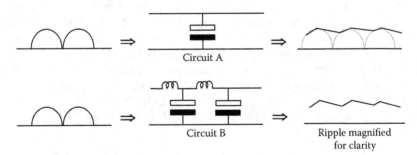

Circuit A

Circuit B   Ripple magnified
                  for clarity

Figure 10.2 Smoothing circuits with results.

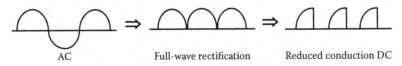

AC   Full-wave rectification   Reduced conduction DC

Figure 10.3 Full and partial conduction of a controlled rectifier.

thus, control can be effectively achieved by adjusting a control potentiometer. Other methods of using thyristors are to put them on the primary side of the circuit or to use them in groups. Both of these methods are uncommon. The thyristor-controlled power supply is in decline, although it is still commonly used for traditional applications and has proved itself to be tough, durable and reasonably energy efficient (up to 80% efficient). It does, however, have some deficiencies. These are a large smoothing circuit is required, the power density is not that large and the control circuit and feedback control is not easily interfaced with modern electronics. For these reasons and because of the greater availability of switch mode power supplies, it is increasingly being supplanted. A traditional thyristor controlled transformer rectifier is shown in Figure 10.4.

10.2.4 Linear

This system is so called because the transistors in the voltage regulator are all working in their linear region. It uses a transformer, a rectifier, a bridge circuit and some smoothing capacitors. It has an electronically controlled voltage regulator at the end of the circuit. This electronic voltage regulator works on a 50 or 60 Hz rectified and smoothed AC. It works by comparing a reference voltage and the output. This error signal controls the output of the regulator, which is a variable electronic resistor where the resistance is very rapidly changed. The output voltage can be controlled using this technique to provide an almost pure DC with a ripple of a few millivolts (Figure 10.5).

Figure 10.4 Conventional transformer rectifier units.

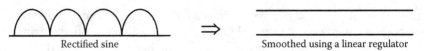

Rectified sine $\Rightarrow$ Smoothed using a linear regulator

Figure 10.5 Effect of a linear voltage regulator on voltage form.

The major disadvantage of a linear power supply is that it is not very efficient, particularly when operating at low voltages. This is because the voltage regulator is operating in a high-resistance mode. This causes substantial amounts of heat to be generated in the voltage regulator, which have to be dissipated. The service life of the electronic voltage regulator very much depends on its temperature. At low temperatures, the linear voltage regulator is very durable. But at approximately 120°C, in continuous use the circuit will eventually be destroyed. Therefore, suitable cooling is necessary for this type of supply, with the maximum output capacity dependent on the heat sink size. This limits the use of this type of power supplies to small zone sizes. It is increasingly being supplanted by small switch mode units.

10.2.5 Switch mode

A switch mode supply is so called because it takes its power input from AC mains power without using a low-frequency (50/60 Hz) isolating transformer to reduce the voltage as do the previously described units. This is the normal situation, although to complicate matters these are secondary switch mode units, which use a step down transformer before acting like a primary switch mode power supply. The secondary switch mode units are becoming rare. The system rectifies 100–240 V AC, and this is passed through a 'chopper' primary switcher to provide a square wave signal between 1 and 200 kHz. This is then passed through a transformer (primary side).

The benefits of this power supply are that the transformer for 200 kHz can be much smaller and more efficient than that used for 50 Hz, and as this is the largest part of a power supply there will be substantial size savings of more than 500% and the same in weight. The secondary side is then again rectified and smoothed. This output voltage is used to control the duty cycle on the primary switch transistor. This gives a relatively smooth output immediately. The advantages of this type of unit are a high compatibility with electronic control and measuring devices, high current output to size ratio, very smooth output (ripple less than 0.1%) and relatively low electromagnetic emissions. Originally, the units are complex with a large amount of components and are consequently likely to be more unreliable; but today they are based on a chip with a few passive components, so they are extremely reliable. As switch mode power supplies have become the standard for televisions, phones, computers and so on, their reliability has been improved so that their mean time to failure (MTF) is normally on the order of 20 years and also their energy efficiency has increased up to a class leading 97%, which provides both a green advantage and as elevated temperatures reduce the life of electronic components a greater lifespan. An example of a switch mode is given in Figure 10.6 and the actual switch mode board and display is given in Figure 10.7.

These units are the most used power supply in general use, and this is likely to become increasingly closely reflected in CP systems, particularly those that use computer control.

Commonly, with a switch mode unit a voltage regulator is incorporated, which initiates a thermal shutdown, that is, the current is switched off when the temperature of the electronics is critically high but still not damaging. When the temperature decreases, the electronics will again act normally. Thermal cut out fuses which blow over at a certain temperature are passing into obselence.

In early switch mode units, mechanical cooling was used with a fan activated over a certain temperature. These are still incorporated on the larger three-phase units, but are not now required with units up to 10 A. An example of a 100 A unit with 16 additional 2 A output zones is shown in Figure 10.8.

The output of a power supply with electronic voltage stabilisation may be adjusted in different ways:

- Manual adjustment of the voltage using a potentiometer.
- Manual adjustment of the current: this can be done by measuring the output current (by passing it through a known resistance) and adjusting the voltage until the output corresponds to the desired current.

Figure 10.6 Switch mode rectifier unit.

Figure 10.7 Switch mode board and display.

Figure 10.8 Switched mode unit with a single high output and 16 small output zones.

- Potentiostatic control: by potentiostatic control, the protection potential voltage of a reference electrode is measured and the current is adjusted via a single comparator until the measured potential corresponds to what is desired. This is not commonly available on the basic systems used for concrete and masonry.
- Control of voltage and current and potentiostatic control using a microprocessor: this is achieved by exchanging the potentiometer with a digital–analogue converter. This interfaces with a microprocessor running an internal program. For current and voltage control, this internal program is sufficient. For control by reference electrode potential, it is normal for the output data of the power supplies and measuring devices to be sent to a personal computer (PC) that runs a software program, which then adjusts the output according to the settings in the program.

10.3 FEATURES OF POWER SUPPLIES

10.3.1 Protection against transients and lightning

Any electrical apparatus connected to the mains supply is exposed to transient surges coming down the supply cables. For instance, these could be caused by lightning. Transients are damaging for most electronic components, which is the reason why it is often necessary to protect power supplies against them.

Power supplies for CP are more exposed than most electrical apparatus as they are often outside, connected to steel reinforcement and anodes. The reinforcement is the most vulnerable part of the circuit as it has a large area and will easily pick up external electrical fields. Anodes on the surface of concrete can be hit by lightning and pass the current down the output cables. Lightning causes big potential differences along cables and consequently a large current flow. Reference electrode measuring circuits are also vulnerable as there is a connection to the steel reinforcement on one side of the circuit.

Thus, there is a need to protect the line input, measuring inputs and direct outputs against transient surge pulses. Depending on the location of the equipment, the need for protection will vary dramatically. Factors such as indoor or outdoor installation, as well as geographical location or heavy machinery levels, should determine the transient protection design. The incidence of lightning is statistically recorded by weather stations and should be considered by the designer so that appropriate measures can be taken. For example, in an area of Japan there have been lightning strikes on a bridge deck CP system on an average of twice a year over several years. They have caused sometimes localised damage and other times complete replacement of power supply cabinets. This has unusually rigorous requirement for protective devices, and these now cost twice as much as the power supplies and data recording equipment.

We have learnt that with a direct lightning strike on a structure there is very little one can do to protect the components in an enclosure (with the surge diverting components in the same cabinet as the power supplies) as the cables carry the energy in, and even if the circuit is disconnected there is widespread arcing damage. If total protection is required, then the protection should be placed in separate enclosures.

The basic principle of transient protection is to drain the transient to ground. It is not possible to stop transients through the use of fuses as the rise time of transients can be very fast and thus they could pass through the circuit before the fuse blows as shown in Figure 10.9.

There are several components that can be used when designing a surge protection system. All of them have various drawbacks and benefits. Often, transient protection is combined with a line filtering function. A description of the commonly used components is given in Sections 10.3.1.1 through 10.3.1.4.

10.3.1.1 Metal oxide varistor

This is a commonly used device because of its low cost and relatively high transient energy absorption capability. It is a non-linear voltage-dependant resistor. Below the threshold voltage the impedance is very high, and over the threshold voltage the impedance decreases and loads the transient. The

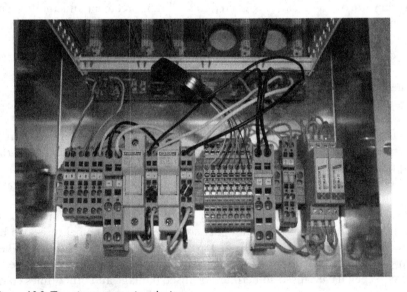

Figure 10.9 Transient protection devices.

drawback of this device is a high slope resistance in the clamping region, which means the clamping voltage is dependent on the current caused by the transient. Another drawback of the varistor is that it ages each time a transient is suppressed. Exposing the varistor to high-energy transients ages it more quickly, and it is not possible to see when the suppressor needs to be replaced.

10.3.1.2 Transient protection diodes

Transient protection diodes are semiconductors and use the avalanche property of semiconductors. They can be uni- or bi-directional for different purposes. Like the varistor, this diode exhibits a non-linear action, but in the clamping region the slope resistance is very much lower. Therefore, the clamping is more effective. If exposed to transients much bigger than those designed for it, it will short-circuit and, therefore, release the circuit breaker or the fuse. The drawbacks of this device are its high cost and comparatively limited ability to pass low current.

10.3.1.3 Surge arrester

A gas-filled surge arrestor comprises a spark gap within a sealed high-pressure inert gas environment. When the striking voltage on the arrester is sufficient, an ionized glow discharge is developed; as the current increases an arc discharge is produced, giving a low impedance path between the electrodes. The arc drop voltage is relatively constant, but the striking voltage to energise it is much higher. The device has a very high current capability but is relatively slow acting. Therefore, it is usually backed up by a fast-acting device. A major drawback is that they tend to remain in the conducting state after the transient has vanished. This requires that a fuse or a circuit breaker is put in series with the surge arrester.

10.3.1.4 Weather station decoupler

These recently developed systems have sensors that monitor the external atmosphere and when it is deemed a high risk of lightning activity disconnects the designated circuits such as the anodes and cathode.

The principal choice for the equipment specifier is to consider how much protection should be specified in view of the cost and likely increase in reliability. Typically on a modular system, there is limited protection on the AC inputs and the positive and negative terminals with no protection on the reference electrode circuits. This is because each of the output and input modules can be simply replaced, and this is cheaper then increasing the protection levels (Figure 10.10).

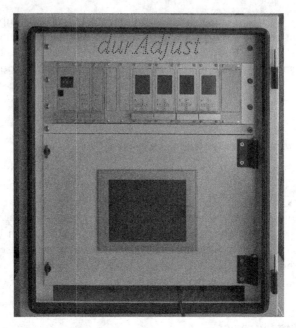

Figure 10.10 Automated power supply system.

10.3.2 Cabinet selection

The cabinet has several functions. Its primary function is to protect the electronics and other electrical items from damage by the environment, and its secondary function is that it should act as a heat exchanger. Its other functions are that it has to be easily opened to permit access, be aesthetically acceptable and provide vandal resistance. Typically, the cabinet is fully enclosed with no cooling vents. It is sometimes possible when placing the unit in an indoor environment to allow cooling vents and thus reduce the installation size. In general, the units are sealed to a protection rating of IP65. This means that the unit is sealed to a level where dust and sprayed water should not penetrate. Normally, glanding is provided by the contractor or electrical specialist on site, and in our experience ingress of the environment will be seen most commonly from the glands and gland plates. Locks are also a problem for sealing. A second common cause of failure is damage to the hinges. It is very important that the hinges are of good quality and have at least the same corrosion resistance as the rest of the cabinet. The best cabinets allow the doors to be removed unhinged when open so that they do not flap in the wind or cause an obstruction. Locks are commonly specified and often prove to be a weak link, both allowing leakage and seizing up.

The international protection (IP) rating guide is given as follows:

SUMMARY OF IP PROTECTION NUMBERS

FIRST NUMBER – PROTECTION AGAINST SOLID OBJECTS

IP tests

0 No protection
1 Protected against solid objects up to 50 mm, for example, accidental touch by hands
2 Protected against solid objects up to 12 mm, for example, fingers
3 Protected against solid objects over 2.5 mm (tools and wires)
4 Protected against solid objects over 1 mm (tools and wires)
   5 Protected against dust-limited ingress (no harmful deposit)
   6 Totally protected against dust

SECOND NUMBER – PROTECTION AGAINST LIQUIDS

IP tests

0 No protection
1 Protection against vertically falling drops of water, for example, condensation
2 Protection against direct sprays of water up to 15° from the vertical
3 Protected against direct sprays of water up to 60° from the vertical
4 Protection against water sprayed from all directions – limited ingress permitted
5 Protected against low-pressure jets of water from all directions – limited ingress permitted
6 Protected against low-pressure jets of water, for example, for use on ship decks – limited ingress permitted
7 Protected against the effect of immersion between 15 cm and 1 m
8 Protects against long periods of immersion under pressure

THIRD NUMBER – PROTECTION AGAINST MECHANICAL IMPACTS (COMMONLY OMITTED)

IP tests

0 No protection
1 Protects against impact of 150 g weight falling from 15 cm height
2 Protected against impact of 250 g weight falling from 15 cm height
3 Protected against impact of 250 g weight falling from 20 cm height
4 Protected against impact of 500 g weight falling from 40 cm height
5 Protected against impact of 1.5 kg weight falling from 40 cm height
6 Protected against impact of 5 kg weight falling from 40 cm height

The cabinet was traditionally sized according to the heat transfer requirements of the electrics; but with the latest high-efficiency switch mode units, this is less important. Generally, the sizing of the cabinet is left to the power supply manufacturer to determine. It is very important for installation and maintenance that there is good spacing for glands and terminals so that in the future maintenance access is possible. This sometimes means that more cabinets should be used with less wires to each unit or the cabinets may have to be enlarged.

The CP designer will normally specify the cabinet material. The advantages and disadvantages of the materials that are commonly available are discussed briefly here.

Glass-reinforced plastic or other filled plastic cabinets are commonly used and have excellent corrosion resistance, especially in saline environments, and are reasonably cheap. Their disadvantages are that the cabinets are prone to damage during transport, have a low vandal resistance, have poor heat transfer and have only fair rigidity, which makes racking systems more difficult. The cabinets are also illegal under the European Union (EU) regulation for electromagnetic interference (EMI).

Mild steel cabinets are normally provided with a polyester powder coating and are tough, have a high heat transfer coefficient, are cheap and allow EMI compatibility. Their problem is that in corrosive environments they start to stain within 1–3 years of installation and after a few more years look aesthetically disastrous. Perforation of the cabinets follows in about 5–10 years, which can be an unacceptably short time period. This propensity can be reduced by maintenance with washing and overcoating. They should be considered suitable for indoor applications only.

Stainless steel cabinets are normally available in two grades, namely, 304 and 316; recently, duplex stainless steel has started being offered. We have found that the 304 grade stains very quickly in a salt-laden atmosphere and the 316 grade is red rust stained within 6 months. The resistance to chloride-induced corrosion of a duplex-grade stainless steel enclosure is likely to be significantly better and might be suitable for long-term use without any additional coating. If the duplex is welded to form the cabinet, then the weld material must be compatible. Stainless steel (possibly apart from duplex) cabinets need to be coated in aggressive environments. Polyester powder coating is the preferred system. If this coating is scratched, red staining may result. Stainless steel cabinets are tough and vandal resistant. They have a heat transfer coefficient between glass reinforced polymer (GRP) and mild steel and allow EMI compliance to be achieved. Duplex cabinets are the most expensive of the cabinets in common use.

The most suitable material for larger cabinets is sheet aluminium coated with a polyester powder coating. These are reasonably priced, have excellent thermal conductivity, are corrosion resistant particularly in chloride-rich environments and are EMI compatible. When the paint film fails, the corrosion product is light coloured and thus staining on the surface is limited.

Some small power supplies use die-cast aluminium boxes, which are powder coated. In most ways, these are ideal as they have excellent heat transmission characteristics, are corrosion resistant, are tough, are cheap and have good EMI compatibility. The only disadvantage is that they are only available in small sizes.

There have been many instances of using cast iron (Lucy) pillars in vandal-prone areas, and these have proved remarkably successful in that they attract little attention and the worst damage encountered has been graffiti. The IP rating of these cabinets is too low to afford satisfactory protection to electrics and electronics for a power supply and additional environmental shielding is required inside this enclosure, so a cabinet within the Lucy pillar is required.

10.3.3 Reading output currents, voltages and potentials

Most power supplies also have the associated reference electrodes terminated in the same enclosure. Thus, there is a requirement for reading the output voltage, output current and reference electrode potentials. The reference electrode potentials are normally required with energised CP systems and 'instant off' systems and after a certain amount of time off.

The simplest, cheapest and crudest way of providing these values is to directly use a portable digital voltmeter on the input and output terminals. The output of the power supply can have a permanent shunt built in to measure the current output. Manually interrupting the current by disconnecting the DC output positive (or switching the power supply off) can then be used to obtain the instant off potentials. This procedure is typical in the traditional CP industry but is not used widely in concrete because of the difficulties in recording data consistently and the relatively low cost of reliable and reproducible meters for voltage, current and potential. With today's technology, it should not be an acceptable design except on a small-scale system or trial.

A modification of this is to use a timer interrupter in the output circuit to provide a more reliable means of obtaining 'instant off' potentials on the reference electrodes. This is advocated in the CEN/TC262 standard, which also advocates the use of a portable meter that is inserted into various sockets in the power supply. The rationale for this poor arrangement is that the portable meter could be quality assured.

It should be explained at this point that to avoid potential drop problems potential measurements are taken as 'instant off' measurements, that is, the potential is measured when the resistive voltage drop from the protection current is gone. This voltage drop occurs as quickly as the current can be switched off, that is, in a few hundredths to tenths of a second. This voltage drop is overlaid onto a potential gradient caused by the protection current having separated the positive ions and the negative ions at the steel interface. Using a portable meter to take 'instant off' measurements with the operator taking an arbitrary first potential seen on the screen after current interruption is not a good or reproducible procedure as in certain locations errors might be on the order of hundreds of millivolts.

To minimise the errors, a system controlled or monitored using potential measurements should be able to be set with a certain time after current interruption until the potential measurement is taken. Ideally, this time delay should be optimised by the corrosion engineer during commissioning and then used for all subsequent measurements. An automatic system like this requires that the control of the various power supplies and reference electrode measurements is coordinated.

Modern power supplies should have direct metering of current and voltage outputs as this dramatically reduces the likelihood of operator error in taking readings and is more user friendly. The choices of meter are moving coil (analogue and now obsolete), light-emitting diode (LED) and backlit liquid crystal display (LCD). These are digital meters, and each one has advantages over the other.

LEDs are accurate and reliable but are difficult to read in direct sunlight. LCDs are accurate and good in light and dark, but have temperature limitations (maximum is normally 60°C). At present, backlit LCD seems to be the best selection for most applications.

The one problem with using integral meters is the fact that they, in common with all meters, need to be calibrated at regular intervals in order to comply with QA testing requirements, unless the reading are categorised 'for information only'. This can be met by the manufacturer providing documentation of the calibration of the meters against a traceable standard and testing the calibration at regular intervals on site. These intervals can be, for example, 5 years as the accuracy required is relatively low (+5 mV) and the meters do not drift, they simply fail.

Automatic measurement of the reference electrode inputs and outputs requires that the system be semi-intelligent. This information is automatically stored in digital form for further future analysis. The next level of sophistication is automatic control systems where a system is controlled using the readings in a dynamic feedback loop according to definable parameters that can be amended by software.

10.3.4 Galvanic separation of power supplies

Concrete structures are often built up in sections, or elements. For several reasons, a structure may have galvanic separation, that is, electrical isolation, between the sections, for instance, in tunnels or other structures with electric rails. This is because induced currents can be very large if the structure is electrically continuous. Another example is separate piers on a marine bridge with precast deck slabs. In either of these cases, if the negative connections within a CP cabinet are made continuous (commoned) there could be substantial current flows, which in certain areas could be much bigger that the CP current. This means that anodic discharges may occur at the steel with the result that corrosion may occur. Induced currents can also play havoc with earthing arrangements.

In the event that parts of a structure are electrically separate, it is desirable to use separate power supplies with separate negative connections for each section or element. This means that a galvanic separation must be maintained throughout the power supply units. To do this, instead of a wired electrical link carrying the communications cable a fibre-optic cable link or a wireless transmitter/receiver is required.

To comply with EMI regulations, everything in the cabinet must be grounded to a common earth. Consequently, the only way to achieve compliance with this EU directive and to have full galvanic separation is to use individual cabinets.

10.3.5 Power supply and reference electrode layout

Normally, during the design of a CP system the engineer makes a decision on the total number of zones required, the number of reference electrodes and the wiring restrictions. These are then divided into areas and the specification of individual control cabinets is then identified with respect to the current and voltage required, the number of outputs and the number of reference electrode inputs. In earlier installations, there was a tendency to put as much electronics into each cabinet as possible; but with the cost of these units falling and the high possibility of errors occurring in the external part of the wiring, that is, outside the power supply enclosure, the modern approach is to have smaller cabinets. These separate cabinets will control a certain number of zones and have the associated monitoring equipment leading back to a central control computer.

The use of junction boxes is very dependent on the form of the structure. If junction boxes are used, they should be carefully specified and sealed as they are a common source of system malfunction. When junction boxes are used, multi-core cable can be run back to the power supply cabinets, reducing the number of glands.

To ease future maintenance, the unit should be constructed in a systematic way using components that are well arranged, easy to operate and as modular as possible. All cables from the CP system should be led into spacious cabinets and terminated in DIN (German standard)-type terminals so that it is easy to undertake measurements with portable tools such as a multimeter. The same cable colourations should be used for the same purpose throughout the power supply, and this should be related to the external wiring:

Red: anode positive
Black: steel (cathode) negative
Blue: reference electrode
Green: communications wire
Orange: buried in concrete

During the initial commissioning of an automatic system, it is simpler if the installation can initially be operated 100% manually. In this mode, it should be possible to connect one anode zone at a time, read the voltage and current from the individual power supplies, measure the current to each zone and measure the potential difference on the individual reference electrodes with the power switched both on and off. As commissioning progresses, the system should be tested in a systematic way for all its automatic functions and this be related to the manual data.

During the operational phase of a CP system, ideally it should be possible to control the installation without using any special instruments apart from a key to unlock the cabinet. Thus, each power supply should have its own identification, a display showing whether the desired current/voltage is being obtained and a screen showing the current and voltage outputs.

10.3.6 Electromagnetic interference

The passage of electrical current creates a magnetic field around a conductor. The electromagnetic field surrounding one electrical device can interfere with another device and cause unintended malfunction. These electromagnetic fields are transmitted by radiated emission or, if the electrical devices are connected to each other by a mains cable, by conducted emission.

All power supplies emit electromagnetic fields, but the amount and the frequency spectrum can be very different. Manual tap transformer rectifiers and linear power supplies generate a little low-frequency conducted emission but do not radiate much energy. Thyristor-controlled power supplies generate a lot of conducted and radiated energy in the low-frequency range. Switched-mode power supplies generate some conducted and radiated energy in the high-frequency range.

A lot of interference can be avoided by careful design of printed circuit boards and filtering at the source. The rest can be shielded with an electrical shield or a Faraday screen.

To ensure correct functioning of electronic equipment, standardisation organisations have specified the maximum EMI that is permitted to be radiated and conducted to the environment by a device. In EU countries, a law from January 1996 was adopted by the member countries. From that date, every device should pass a test in which the maximum EMI it is permitted to radiate and conduct is specified and the minimum limit of EMI it must be capable of resisting is also prescribed. When the test has been passed, the equipment is allowed to carry a 'Certificate Europe' (CE) label. All electronic equipment must carry the CE label when sold in the EU market.

This legislation is divided into two environmental classes. Part 1 is for electronic devices in residential, commercial and light industry applications, and part 2 is for industrial environments. There is now the European standard called EN61000-6-3-3:2008 electromagnetic compatibility. Within this are the older standards for emission (EN50082-1:1992 electromagnetic compatibility) and immunity (EN50082-2:1994 electromagnetic compatibility).

10.3.7 Efficiency

Normally the consumption of electricity on a concrete CP system is not of great importance to the overall running cost of an installation and, thus, the total efficiency of a power supply is of interest only in the amount of heat that has to be dissipated in an individual cabinet. In some countries, however, there is an 'environmental audit' to minimise the consumption of resources, and in these locations high-efficiency power supplies may be required.

Similarly, in the unusual occasion where mains electricity is not available and solar, wind or other power is required the efficiency of the power supplies should be maximised. When this is done, a realistic estimate of the actual power required (driving voltage and current to give wattage) should be made. In this case, the anode should be selected to ensure the lowest possible driving voltage required.

10.4 AUTOMATIC SYSTEMS

For the past 15–20 years, it has been common to automate the control and operation of CP systems. This trend has become increasingly apparent as the relative cost of these systems has dropped and their capabilities and reliability have improved.

It is now almost universal to automate the CP system on a large structure with a large number of zones with individual power supplies and reference cell inputs. For each zone, it is normal to actively control the maximum voltage, the maximum and minimum currents, the optimum potential and the maximum or minimum depolarisation.

An intelligent system will actively control the outputs according to feedback loops similar to the adjustments made by an experienced CP engineer while storing all aspects of the system performance. This system should also be able to adjust for failures, for example, buried reference electrodes sometimes start giving 'spurious' readings, which could be ignored if sufficiently deviant. Other failures of the anode, cathode connections or power supplies should be recorded with readjustment to accommodate these problems in the most efficient way possible.

When specifying a system that is automatically controlled and remotely operated, there are several factors that should be considered to arrive at the best solution for the project. The first and most important consideration is to decide what you want the automatic system to do. The next consideration is that if a limited specification is given then the contractor is going to propose the cheapest system when there is a limited number of manufacturers with a wide range in complexity. This consideration leads some consultants to specify the units, but many are unwilling to take this responsibility.

The normal practice is for the specifying engineer to specify the output size and number and the number of inputs, then outline what he or she wants the system to do and then specify some recommended generic specifications that it should be in compliance with. This arrangement absolves the designer from specifying a single system and allows several competing systems to be proposed. The principal matter for the designer then is to decide what he or she specifies the system to do without either excluding the more desirable systems available or forcing the contractor to follow the lowest-cost route. This is a difficult balancing act. A further complication is that it is necessary that the designer keeps a firm hand on reality at this point as non-standard individual additional features cost a lot of money to develop dedicated software or hardware and adds to the unreliability of the control system. An intelligent system should ideally give information and control on several levels:

- If the system is actively controlling using data from reference electrodes (or other monitoring equipment) to provide optimal protection levels with minimal anode damage: to do this, an automatic system needs some form of data storage facility; a computer processor; and a way of interfacing with potential, current and voltage inputs. This in practice requires that the system be based on some form of PC or programmable logic controller (PLC) technology.
- The system should be simple to operate and allow the operating parameters to be changed easily.
- Historic operational data should be directly accessible in a simple form. Ideally, this should be available in Microsoft Excel files.
- Is the installation functioning correctly? Yes/no.

- If no, what is occurring to prevent the installation from functioning correctly (i.e. hardware damage, data communications damage or software crash)? Also, when and where did the problem arise?
- It should be possible to copy data to another computer where a full record of the operation of the system can be examined in as detailed a manner as required.

The system should be constructed in the most modular way possible to permit simple servicing and maintenance.

To undertake this task, the automatic system has to have several components, and these are outlined here:

- Modem: to transmit data and upload commands and updated software. This should be robust and able to work at the maximum speed of the telephone lines to minimise the communication time. Wireless Global Standard for Mobile Communications modems have become quite popular, but can have poor reception and low data transmission rate. More recently, there has been a trend towards using wireless routers, which communicate with the remote system through the Internet. Some recent systems are uploading and storing data on 'clouds'.
- Control unit: it is usual to use PC or PLC technology as they are common and cheap and powerful. In the event that these units are used, 'industrial-quality' components should be used to give higher reliability and future compatibility.
- Data storage facility: this is a hard disk or flash random access memory (RAM). Flash RAM is more expensive, but should be specified due to its much greater life expectancy. This is particularly important as unlike an office PC the systems operate for 24 hours every day in varying temperatures.
- Input analogue to digital convertors: there are several systems on the market. These preferably should be designed for industrial use and have as high noise rejection levels as practically possible. They should also have a high input impedance. The best examples are in the gigaohm range. It is correct (to avoid drainage effects from adjacent zones) requirement that all the potential values from the reference electrodes on the system are taken with the current output of anodes simultaneously disconnected. To obtain these 'instant off' values, holding circuits are required where the information is captured at the predetermined time and then released in series through the digital system to the controller.
- Software: this is probably the most critical part of the unit. The program should be stable and reliable and written to assist a corrosion engineer in running the system. The system should be simple to operate and make it simple to record and display the data.

- Operations point: the system installed has to be interrogated on site. In the latest examples, this is normally undertaken with an integral touch screen. For more complex interaction, a keyboard can normally be coupled to the computer. Some of the older systems had specially installed switches around the LCD screen.
- Communications between the various cabinets: This allows data to be passed through the system. They normally conform to an industry standard such as RS485 and can be hard-wired, fibre optic or wireless radio. Hard-wired is the cheapest and most common, but the other two systems also are in widespread use.

A modern automatic system can easily store a vast quantity of data. By storing data as integrals in binary form, there are no practical problems in recording literally millions of values. If there is too much data, then there is a big risk that the collation of data becomes too troublesome for the user relative to the benefit and it is ignored. To try and avoid this, data compression should be used early in the system to try and filter the relevant information from other data that is not practically relevant.

Data may be compressed by the following techniques:

- Saving the average value and the standard deviation over a suitable period: this, for instance, could be 24 hours if no essential variations in the environment are occurring; every 3 hours if there is a small tide influence; or 1 hour if there is a large influence by concrete temperature, rain, tidal water or air temperature.
- Only saving data when there are considerable changes relative to the last data set saved.

Data compression showing the 24 hour average data from a bridge pile.

A modern automatic system typically comprises several cabinets at spaced out locations around the structure with a master control cabinet, which each of these units communicates with. The master control cabinet is normally a communications point allowing remote access. Data are sent from the cabinets to the master in one of the following four ways:

1. Communications wire, which is normally a single or twin twisted pair: this is the most common arrangement and has several advantages in that it is cheap, simple to wire and reliable. It has a few disadvantages: it does not cope with long distances well and normally needs to be boosted at 1000 m intervals. Other disadvantages are that it can disrupt galvanic isolation and is vulnerable to transients. Finally, the wiring itself may be difficult or expensive to undertake.

2. Fibre-optic communications wires: these cables allow interference-free communication over longer distances (tens of kilometres) and galvanic isolation between cabinets. Its disadvantage is cost and the difficulty of making connections, but the cost difference is getting less.
3. Radio or microwave: radio systems commonly use a free frequency where a licence is not required. The combined transmitter/receiver is mounted in each (or a group) of cabinets, and this communicates with the master controller. Ranges of over 5 km are achievable with this system, but aerials need to be carefully positioned and there is a significant chance of interference.

Chapter 11

Monitoring cathodic protection in concrete and masonry structures

John Broomfield

CONTENTS

11.1 INTRODUCTION

Monitoring cathodic protection systems require a number of elements. There are the probes that do the monitoring, the monitoring system that operates the probes and collects the data, and the performance criteria against which the data is assessed. This chapter describes the probes, the monitoring systems, the control criteria and regimes and the requirements of the major international standards for monitoring and control of cathodic protection systems for stele in concrete. A brief discussion of steel-framed masonry structures is also included.

11.2 MONITORING IMPRESSED CURRENT CATHODIC PROTECTION SYSTEMS

There are a number of different monitoring probes that can be used to ensure that a cathodic protection system is functioning correctly, that is protecting the steel at the lowest current and voltage levels to minimise durability problems. These must be wired to a control and monitoring system with a suitable regime of manual or automatic data recording and reporting.

11.2.1 Monitoring probes

There are a range of probes that can be used to monitor cathodic protection systems. The most important of these is the reference electrode or half-cell, as it is referred to in all the major standards on cathodic protection of steel in concrete. Other probes may be used to provide complementary information, or to substitute for reference electrodes in some circumstances.

11.2.1.1 Reference electrodes

A reference electrode is a metal in a standard solution of its own ions, which gives a fixed and reversible potential under controlled conditions. According to the European Standard on cathodic protection of steel in concrete, BS EN 13696: 2000, suitable reference electrodes for embedding in concrete include Ag/AgCl/0.5 M KCl gel electrodes and $Mn/MnO_2/0.5$ M NaOH electrodes. The Ag/AgCl electrode is widely manufactured in Europe and North America. The MnO_2 electrode is proprietary to a European manufacturer.

Present designs of reference electrodes for reinforced concrete applications are very robust. The main problems arise during installation. Shrinkage of the cementitious grout around the porous electrode tip may occur, resulting in poor contact with the parent concrete. This situation can result in readings that are unstable or inexistent. This can usually be identified and rectified during the commissioning or maintenance period. The other major problem is that signal cables can be damaged during the installation process. Once installed and operating properly, reference electrodes are very reliable and survive in working order throughout their specified life unless exposed to extremely harsh environments (such as prolonged drying or freezing), contamination or physical damage.

The life of a reference electrode is finite, depending on the mass of silver metal and the current passed, which depletes it, that is, number and length or reading time. This assumes no other deterioration mechanism. If an installation is designed for a very long life and reference electrodes are inaccessible, pseudo-reference electrodes may be installed to supplement the 'true' reference electrodes.

11.2.1.2 Pseudo-reference electrodes or potential decay probes

Electrodes of graphite, mixed metal oxide (MMO) coated titanium and zinc have been used [1]. Lead has also been used, but only in electrochemical realkalisation and chloride removal systems as far as this author is aware. Zinc is used in general seawater applications and has been widely used for underwater portions of reinforced concrete bridges or jetties when it is immersed directly in the sea. However, zinc is less stable in concrete where the chloride and oxygen levels fluctuate and the corrosion products cannot be washed away. Graphite and MMO are also sensitive to the level of oxygen which can change dramatically if concrete saturates and dries out.

As discussed in Chapter 4, the monitoring criteria for impressed current cathodic protection of steel in concrete require the measurement of an electrical potential difference over a period. If the pseudo-reference electrode is stable over that period, compared to an adjacent true reference electrode, its reading can be considered valid. The installer is advised to check graphite electrodes against adjacent embedded Ag/AgCl/KCl electrodes during commissioning to determine their calibration and stability.

11.2.1.3 Luggin probes

Luggin probes are not probes in their own right but a 'bridge' of low electrical resistance from the surface to the embedded steel to allow a portable reference electrode to be used to monitor the steel potential despite the presence of an anode between the surface and the steel. It also minimises the effect of the 'IR' drop, discussed in Section 11.2.3.1.

Although they are not widely used in commercial installations, they have their uses in trials and in areas that need monitoring but are either awkward for permanent reference electrode installation or where the cost of installing many electrodes, wires and monitoring becomes excessive and manual positioning of a portable electrode is feasible.

11.2.1.4 Null probe

The 'null' probe was used for a while in North America before the performance of reference electrodes became highly reliable. It uses a piece of electrically isolated steel, ideally a piece of reinforcement in the most anodic area. Electrical connections are made to the reinforcement and to the isolated steel and connected to an ammeter. If the steel is sufficiently anodic, corrosion current will flow from the anodic null probe, through the wires, to the steel. As an increasing level of cathodic protection is applied, the current reduces and then reverses as the potential is depressed to that of the surrounding steel [2] (Figure 11.1).

11.2.1.5 Resistance probes

A resistance probe is a proprietary probe made from a thin sheet of steel, sometimes in the form of a tube when used in concrete to simulate a reinforcing bar. It is embedded in the concrete either from new (a cathodic prevention system) or in a grout with a similar or higher chloride content than the original concrete for an existing, corroding reinforced concrete structure. If the probe corrodes, the thinning of the steel gives a higher electrical resistance. This can be accurately measured with temperature compensation using a Wheatstone bridge circuit. The probes work best in a flowing electrolyte where the thinning of the probe is uniform. The pitting attack on steel seen in concrete tends to lead to rapid failure of resistance probes.

Resistance probes as best used when they can be built into new concrete when they will see the same environment as the rest of the steel. They can be useful as a simple 'go/no go' showing that cathodic protection is protecting the steel.

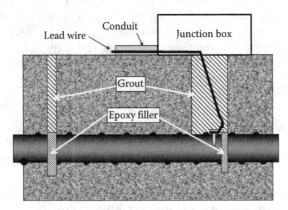

Figure 11.1 A null probe design. (From Bennett, J. E. and Broomfield, J. P. *Mater. Perform.* 1997 Dec, 36(12), 16–21.)

11.2.1.6 Other probes

Simple coupons can be cast into concrete to measure current flow. Like the resistance probes, these suffer from being in a grout and therefore not in the same conditions as the actual reinforcing steel unless cast in at time of construction.

11.2.2 Monitoring equipment

All systems that conform to NACE, European or related standards will consist of a direct current (DC) power supply for each anode zone, a set of reference electrodes embedded in each zone all terminated back at a control board, along with connections to the reinforcement for each anode zone and each reference electrode or set of reference electrodes.

The details of transformer rectifier power supplied are discussed in Chapter 10.

The power supply may be run in two different modes:

1. At constant current with the voltage allowed to fluctuate within the limits of the power supply or narrower present limits
2. At constant voltage with the current allowed to fluctuate within the limits of the power supply or narrower present limits

It may also be run at a constant potential measured against a single reference electrode. This is more common in fully immersed zones.

If there is more than one reference electrode per zone, it may be difficult to achieve the required 100 mV depolarization on all reference electrodes as one may be in a far more anodic location than another. An engineering judgement is then required as to how to 'balance' the requirements of the system.

The cathodic protection engineer needs to carry out the following actions when carrying out the adjustment and monitoring requirements of BS EN ISO 12696:2012 or NACE RP0290 [3]:

1. Measure the current and voltage in the system
2. Measure the 'instant off' potential of the steel versus the reference electrodes
3. Undertake a depolarisation test on each zone to ensure that that they are all achieving 100 mV depolarization, recording the instant off and then the depolarised potentials at a number of time intervals
4. Plot a graph of depolarisation versus time
5. Adjust the current and/or voltage
6. Report his/her results and actions to client

He/she may do this on site with a manual system or a system with a logger built in. Alternatively, he/she may do it from his/her office with a remote monitoring system. However, even remote monitoring systems require one or two site visits per year for a visual check that everything is intact and performing properly.

11.2.2.1 Basic monitoring systems

The most basic system will consist of a termination box containing terminations that can be connected to via a high impedance digital voltmeter. This is used to take readings manually. The outputs from the DC power supplies are adjusted manually, as shown in Figure 11.2.

11.2.2.2 Logging

A logging system will include a logger to store data of current, voltage and reference electrode potentials. It may include a facility to record instant off as well as 'on' reference electrode potentials. Data are usually collected by visiting site and downloading them to a laptop computer or other suitable digital storage device.

Figure 11.2 A simple manual impressed current cathodic protection power supply with four zones adjusted by the potentiostats at the top of the cabinet.

11.2.2.3 Automated or remote monitoring systems

An automated system will normally be remotely accessed by a telephone line or similar suitable communications system. Satellite communications have been used in remote locations. The system can be adjusted remotely. In some cases, the system is capable of carrying out the commissioning process automatically according to BS EN ISO 12696:2012 or NACE RP0290 [3]. The software may allow for alarm levels to be set and for alarms to be sent to the host computer, via the Internet, to pagers or to other communication devices.

Alarms might include the following:

- Power going off
- Power coming back on
- Instant off potentials exceeding (becoming more negative than) present levels such as those in BSEN13696 for hydrogen evolution
- Instant off potentials being more positive than present levels (such as the original rest potential, implying that the steel is not polarising)
- Current reaching an upper limit or voltage reaching a lower limit (could indicate a short circuit)
- Current dropping to a lower limit or voltage reaching a higher limit (could indicate excessive drying out or a break in the circuit or could exceed the pitting potential of titanium wires) (discussed later in this section)

Figure 11.3 shows a screen from a remote monitoring system for a cathodic protection system on a set of bridge piers in the United Kingdom. By clicking on the relevant part of the screen, the cathodic protection engineer can do the following:

- Pull up schematic or computer aided design (CAD) drawings of the structure showing zones, reference electrode locations and current instant off measurements
- Read the current and voltage to each zone
- Read the historic data
- Set up a depolarisation test
- Read previous depolarisation test results and view graphs

11.2.3 Criteria

The most important part of the monitoring process is ensuring that sufficient current is being passed to control the corrosion process, while ensuring that the current and voltage are not excessive. A well-designed system will have sufficient reference electrodes located at representative locations to ensure that the measurements indicate the overall behaviour of the structure. Excessive current will consume the anodes more rapidly than

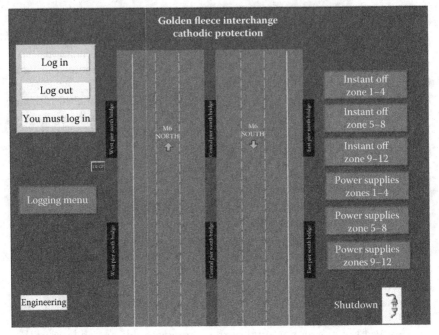

Figure 11.3 Main operational screen for a remote monitoring system for a motorway bridge in the United Kingdom. Courtesy of ElectroTechCP Ltd., Grantham, United Kingdom.

necessary and degrade the concrete anode interface by the acidification of the anodic reaction.

11.2.3.1 Polarisation/depolarisation and the IR drop

The major standards propose potential decay criteria as the most suitable for atmospherically exposed reinforced concrete structures. The potential decay value is measured either from instant off to a given period, typically up to 24 hours, or over a longer period.

The instant off requirement recognises the fact that the reference electrode is located in an electric field formed by the current flowing through the resistive concrete. This is shown in Figure 11.4. To measure the electrochemical potential between the steel in concrete and the reference electrode more accurately, it must be done at the instant of switching off the cathodic protection current. In practice, the measurement is made a few tenths of a second after switch off because of capacitance and hysterysis effects. The on potential will frequently differ from the instant off potential by a few millivolts to hundreds of millivolts. In highly resistive media such as steel-framed structures in brick or stone work, the IR drop may be several volts.

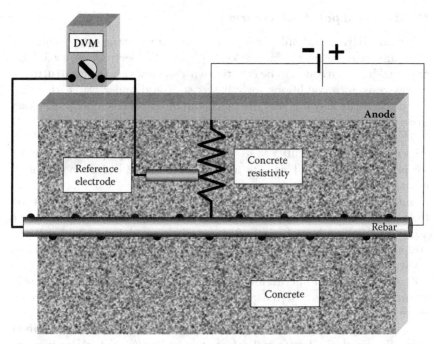

Figure 11.4 The IR drop effect. When the cathodic protection is on, the digital voltmeter (DVM) registers a voltage proportional to the impressed current and the location of the reference electrode relative to the anode and rebar.

The most commonly adopted 100 mV decay criterion is an empirical one based on long experience.

The advantages of the polarisation/depolarisation criteria are as follows:

1. The criteria measure changes in potential so the absolute calibration of the reference electrode is not essential, allowing for drift in embedded sensors over time
2. The requirement to measure the potential over time ensures that instabilities in the system are noted and rectified

The disadvantages of the criteria are as follows:

1. The measurement takes 4–24 hours
2. Changes in the environment can affect the reading (e.g. because of tidal movement and heavy rain)
3. In some cases, depolarisation can take a very long time, typically because of oxygen starvation at the steel surface. In such cases alternative criteria may be used.

11.2.3.2 Fixed potentials criteria

For many buried and submerged structures, an instant off potential of −720 mV with respect to a 0.5 M Ag/AgCl/KCl electrode is used. This potential has been found to be effective with pipelines and other structures. In low-resistance conditions such as wet soils and under water, the measured potentials may not vary too much and the depolarisation may be very slow. This criterion may be useful in such conditions.

The CEN standard requires all potentials to be kept above −1100 mV with respect to Ag/AgCl/0.5 M KCl for plain reinforcing steel and −900 mV for pre-stressing steel. This is to minimise the risk of hydrogen embrittlement, particularly of pre-stressing steel.

Since the CEN standard requires all potentials to be more positive than −1100 mV and the absolute potential criterion is more negative than −720 mV, the window of −380 mV for unprestressed and 180 mV for pre-stressed steel is very narrow given the wide variations in potentials found on atmospherically exposed reinforced concrete structures. This is why the 100 mV depolarisation criterion is usually preferred except for concrete in saturated conditions where depolarisation is very slow and potentials are more uniform.

In a note, the CEN standard refers to 'an investigation criterion' where the fully depolarised potential (after at least 7 days off) is less negative than −150 mV with respect to Ag/AgCl/0.5 M KCl. However, this assumes the validity of the ASTMC876 criterion that says that there is a low probability of corrosion at such potentials. This may be valid for chloride contaminated steel in concrete. This author recently found corrosion rates of 3–5 μm section loss per year on a steel-framed brick clad building with steel potentials less than −100 mV with respect to Ag/AgCl/0.5 M KCl. Before using such a criterion the engineer needs to understand why the other criteria are not being achieved and be certain that −150 mV with respect to Ag/AgCl/0.5 M KCl is a valid 'non-corroding' criterion.

It should be noted that basic electrochemistry tells us that it takes a smaller amount of current to shift the potential of a piece of steel that is not corroding than for a piece of steel that is corroding. This is embodied in the Stern and Geary equation [4], which can be written as

$$I_{corr} = K \cdot \frac{\Delta I}{\Delta E}$$

where I_{corr} = the corrosion current

  K = a constant
  ΔI = an applied current
  ΔE = the shift of the potential

The equation is strictly valid only for a shift of around $\pm$ 10 mV from the rest potential, but the principle is evident. If the corrosion rate I_{corr} is high, then it will take a higher applied cathodic protection current ΔI to achieve a 100 mV shift ΔE than if I_{corr} is low.

If the potential is not shifting in a non-corroding area, it implies that current is not reaching the steel. This may be due to a problem with design of the anode layout, installation or performance of the anodes or the electrical resistance between the anode and the steel.

11.2.3.3 E-logI plots

The plotting of the instant off potential of the steel versus current on a logarithmic (current) scale may show a 'break point', which in principle is the change over from anodic to cathodic conditions. However, the graph is often a gradual curve with no clear break point. The shape of the curve changes depending upon the sweep rate. This technique was described in some detail in the original version of NACE RP0290, but is only mentioned in passing in the current version as a method of determining the initial current setting. It is not mentioned in BSEN13696.

11.2.3.4 Current density requirements and criteria

In a survey of 287 North American bridge CP systems where de-icing salts were periodically used, it was noted that most protection current densities should have been below 5.4 mA·m^2 after an unspecified initial period of protection [6]. These observations and those of others suggest that a long-term protection current density of 5 mA·m^2 or less may be sufficient to prevent corrosion initiation [7].

It was also suggested [5] from laboratory tests that constant current densities could be used if the chloride concentration at the reinforcement was reliably known according to Table 11.1.

However, it was noted that these suggested current densities were based on fairly short-term laboratory testing and would need further investigation.

The CEN standard gives no criteria referring to current densities but in the non-mandatory design appendix suggests that systems should be

Table 11.1 Experimental current density criteria

Chloride by mass of cement	Chloride by mass of sample	Current needed (mA/m²)
<0.2	<0.6	0
0.2–0.3	0.6–1.2	5
0.3–0.8	1.2–13.0	13
0.8–1.6	13.0–6.0	17

Source: From Bennett, J. E. and Broomfield, J. P., *Mater. Performance*, 1997 Dec, 36(12), 16–21.

designed for current densities up to 20 mA·m^2 for actively corroding systems and 0.2–2 mA·m^2 for non-corroding (cathodic prevention) systems.

In this author's experience, most systems will achieve the 100 mV criteria at 5 mA·m^2 or less. Very few require more than 10 mA·m^2 with the exception of some systems in very aggressive environments such as those found in the Arabian Gulf region and a few very highly chloride contaminated bridge decks in North America. Many systems are overdesigned in terms of their design current which makes them difficult to adjust, inefficient to run and potentially reduce the anode life if the current density is set too high. Examples of current and potential data are given in Tables 11.2 and 11.3.

It is important to recognise that current demand drops with time as chlorides move away from the reinforcement, the passive layer is re-established and benign reactions occur at the cathodic steel surface. Even in aggressive environments, the current level required to achieve the protection criteria will decline. If the environment has been improved, for example by draining saline water away from the concrete surface, this effect will be more rapid and more noticeable.

11.2.3.5 Voltage limits

Some anode systems, known as discrete or probe anodes, rely on a titanium wire connecting the anodes in strings forming zones or sub-zones. In the presence of chlorides, bare titanium can pit. The pitting potential for titanium in a chloride solution refers to the potential between the wire and the reinforcing steel which may be far lower than the voltage difference measured between the positive and negative terminals on the power supply. The manufacturer's recommendations are usually to keep the zone voltage below

Table 11.2 Preliminary data

Location	
Structure	Tower
Zone description	Zone 1
Zone location	Tank room
Date zone energised	2003/01/13/13:00
Location of commissioning file/report	C:\XXX\AAAA
Location of quarterly reports	C:\XXX\BBBB
Location of annual reports	C:\XXX\CCCC
Date of depolarisation	April 15, 2003
Time of depolarisation	13:00
Current at depolarisation (mA)	1500
Current density at depolarisation (mA)	5
Voltage at depolarisation (V)	5.5

Table 11.3 Potential data

	Potentials (Ag/AgCl/0.5 M KCl)				Depolarisation			
	RE1 (V)	RE2 (V)	RE3 (V)	RE4 (V)	RE1 (V)	RE2 (V)	RE3 (V)	RE4 (V)
Averaged on	−0.840	−0.745	−0.414	−0.375	0.218	0.142	0.055	0.077
Instant off	−0.621	−0.604	−0.360	−0.298	**IR Drop**			
1 hour	−0.186	−0.109	−0.325	−0.188	0.435	0.495	0.035	0.110
4 hours	−0.058	−0.008	−0.239	−0.100	0.563	0.596	0.131	0.198
8 hours	−0.058	−0.008	−0.239	−0.100	0.563	0.596	0.131	0.198
24 hours	−0.007	−0.005	−0.031	−0.011	0.614	0.598	0.329	0.287

Stability	**RE1**				**RE2**			
	Ave	Std	Max	Min	Ave	Std	Max	Min
On 08:00–13:00	−0.840	0.082	−0.621	−0.256	−0.745	0.053	−0.604	−0.165
20 ± 4 hours	−0.013	0.003	−0.007	−0.009	−0.006	0.000	−0.005	−0.001

Stability	**RE3**				**RE4**			
	Ave	Std	Max	Min	Ave	Std	Max	Min
On 08:00–13:00	−0.414	0.021	−0.360	−0.069	−0.375	0.029	−0.298	−0.091
20 ± 4 hours	−0.048	0.013	−0.031	−0.039	−0.013	0.001	−0.011	−0.002

8 V DC. This is an important limitation especially when dealing with small high-resistance zones, such as for steel-framed masonry-clad structures.

11.2.4 What the CEN and NACE standards say

The NACE and CEN standards on cathodic protection have a lot of similarities and a few differences:

1. Both standards refer to instant off potential measurements being made between the reference electrode and the steel 0.1 to 1.0 second after switching off the DC circuit.
2. NACE RP0290 refers to 100 mV polarisation development, from rest to instant off as well as 100 mV depolarisation from instant off to a period afterwards.
3. NACE does not refer to a period over which the 100 mV difference should develop. EN13696 refers to 100 mV depolarisation over 24 hours and 150 mV depolarisation over a longer period.

4. The CEN standard refers to 'any representative spot' meeting the criteria.
5. NACE says the criteria should be applied 'at the most anodic location in each 50 m² area or zone, or at artificially constructed sites'.
6. The CEN standard requires all potentials to be kept below −1100 mV with respect to Ag/AgCl/0.5 M KCl for plain reinforcing steel and −900 mV for pre-stressing steel.
7. The CEN standard refers to representative points meeting 'any' of the depolarisation criteria of 100 and 150 mV or an absolute (instant off) potential of −720 mV with respect to Ag/AgCl/0.5 M KCl.
8. The NACE standard refers to an E-logI criterion or set up criterion. However, little information is given on how to create and interpret the E versus logI curve.
9. Both standards refer to the need for stability over the period of measurement.
10. The CEN standard recommends avoiding taking reference electrode potential measurements within 0.5 m of repairs. The NACE standard says that the original concrete around the steel should not be disturbed during reference electrode installation. However, see item 5.
11. In a note, the CEN standard refers to an investigation criterion where the fully depolarised potential (after at least 7 days off) is less negative than −150 mV with respect to Ag/AgCl/0.5 M KCl. However, this assumes the validity of the ASTMC876 criterion that says that there is a low probability of corrosion at such potentials. This may be valid for chloride-contaminated steel in concrete. However, this author recently found corrosion rates of 3–5 μm section loss per year on a steel-framed brick clad building with steel potentials less than −100 mV with respect to Ag/AgCl/0.5 M KCl.

The CEN standard gives a detailed maintenance and monitoring procedure. Routine inspection procedures shall be as follows:

1. Functional check of the following:
   a. Confirmation that all systems are functioning
   b. Measurement of output voltage and current to each zone of the cathodic protection system
   c. Assessment of data
2. Performance assessment of the following:
   a. Measurement of 'instantaneous off' polarised potentials
   b. Measurement of potential decay
   c. Measurement of parameters from any other sensors installed as part of the performance monitoring system
   d. A full visual inspection of the cathodic protection system
   e. Assessment of data
   f. Adjustment of current output

It states that the function check should be undertaken monthly initially extending to quarterly. The performance assessment should be annual. The NACE standard is in agreement.

11.2.5 An ideal monitoring program/report for an impressed current cathodic protection system

An ideal monitoring report for a depolarisation test would give the following information:

- Date and time of monitoring
- Weather conditions
- Damage or changes in the system since last report
- Current and voltage levels prior to depolarisation test
- Clear information on whether the system is being run at constant current, constant voltage or constant potential
- Stability of on and instant off potentials over a reasonable number of measurements prior to test
- Depolarisation curves of all reference electrodes in each zone
- Table of depolarisation data
- Interpretation as to which reference electrodes are meeting the criteria
- Information on the system as left

11.2.6 Steel-framed masonry structures

Steel-framed masonry structures (brick or stone clad), can be cathodically protected if there is damage to the masonry because of expansive corrosion of the steel frame. This is an increasing problem with early twentieth century buildings, some of which are of architectural and historic importance. There are several important issues when addressing such structures that are not relevant to this chapter and are discussed elsewhere in this book. However, the monitoring and control of such systems are relevant.

The 100 mV polarisation and depolarisation criteria are most relevant to such structures. The structural steel is usually highly exposed to the atmosphere, so depolarisation can be very rapid and the IR drop effect can be very big as the electrical resistance of mason is far higher than concrete, typically in the region 100 kΩ·cm compared to 5–50 kΩ·cm for concrete that will support reinforcement corrosion.

Often the zone size is small because steel columns and beams are of very different dimensions. Also the current demand is small as some areas of steel may not be in contact with the mortar and masonry. This makes design of the system more complicated, as small zones with small current densities need higher voltages to drive them. Many of these systems use probe anodes with titanium wire connectors, which limits the voltage output.

11.2.7 Common problems and troubleshooting

Figure 11.5 is an example of instability in the reference electrodes in zone 1 of a system. The reason can be seen from the current and voltage plot, Figure 11.6, which shows instability in the DC and voltage outputs.

The reason for the instability was not fully identified but was due to some form of electrical interference. The addition of smoothing circuits to the power supply eliminated the problem. This was not an ideal solution but was the most cost effective.

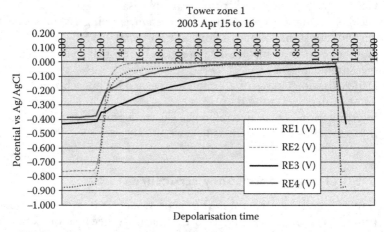

Figure 11.5 Depolarisation curves for a four reference electrode zone.

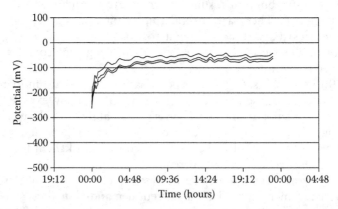

Figure 11.6 Unstable depolarisation plots of zone 1 of a motorway bridge impressed current cathodic protection system.

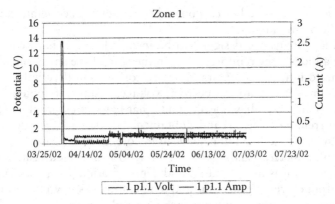

Figure 11.7 The current and voltage plots of zone 1 of a motorway bridge impressed current cathodic protection system.

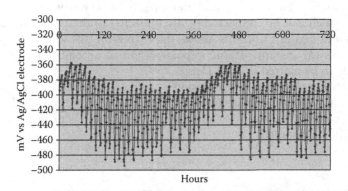

Figure 11.8 Hourly potential data for reference electrode 1 zone 1 (Pier 1 Tidal Zone).

Figures 11.7 and 11.8 graphically illustrate the effect of tidal movement on potentials. The potential shift is up to 140 mV between high and low tide. This effect can make it very difficult to measure depolarisation because of variation in conditions.

11.3 MONITORING GALVANIC (SACRIFICIAL) SYSTEMS

Galvanic cathodic protection systems are becoming more popular. In some cases, they are seen as lower maintenance options than impressed current cathodic protection. In other cases, they are being used to enhance patch repairs that would otherwise risk suffering from the 'incipient anode' or 'ring anode' effect where the creation of a cathode (the steel in the chloride free repair) where previously there was an anode (the corroding rebar in chloride contaminated or carbonated concrete) can accelerate corrosion around the original repair.

Galvanic systems are designed to be installed with a hard connection from the anode to the reinforcement. In the case of the small zinc disks in mortar fitted into patch repairs, the wire connection directly to the reinforcement makes monitoring current, voltage, instant off potentials and depolarisation impossible. Specific units that have an external wiring system for current monitoring can be obtained. The design of anode in which they are strung together and embedded in core holes are easier to connect though an ammeter, usually with a logger, to measure current. A switch can be used to measure instant off and depolarisation potentials. A plot of anode output current versus time and temperature is shown in Figure 11.9. The strong correlation between temperature and current output can be seen (Figure 11.10).

The thermal sprayed zinc based anodes are connected directly to the reinforcement network via a stud and a plate. However 'windows' can be created which can be connected or disconnected to the rest of the anode, as shown in Figure 11.9. This allows the current in that area to be logged and allows the instant off and depolarisation of an embedded reference electrode to be plotted (Figure 11.11).

In such installations, less attention is paid to the risk of short circuits between anode and reinforcement as they are directly coupled. However, tie wire, shallow steel or other short circuits must not occur in areas where monitoring is installed.

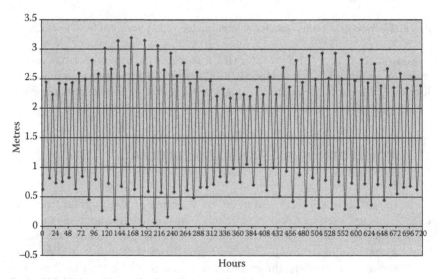

Figure 11.9 High and low tide chart from nearby location.

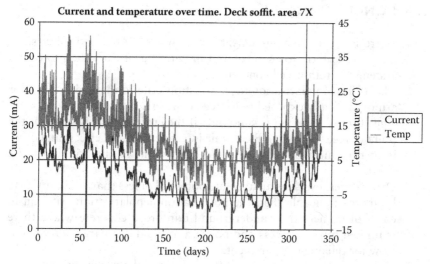

Figure 11.10 Graph of galvanic current and temperature versus time for bridge soffit. Courtesy of Fosroc Ltd., Birmingham, United Kingdom.

Figure 11.11 A window for current and reference electrode potential monitoring in a Corrospray™ Al-Zn-In thermal sprayed anode system on a marine bridge. (From Daily, S. F. and Green, W., Galvanic cathodic protection of reinforced and pre-stressed concrete using CORROSPRAY™ a thermally sprayed aluminium alloy, Corrpro technical Library paper CP-51. 2007.)

11.4 CONCLUSIONS

1. There is reasonable agreement between the NACE and CEN standards on criteria for control and monitoring of cathodic protection systems for reinforced concrete.
2. The 100 mV polarisation/depolarisation criterion is widely used for atmospherically exposed reinforced concrete.
3. The 100 mV criterion is valid only if the environment is stable at the point of measurements. Electronic interference, tidal effects and other influences must be identified and dealt with for the criterion to be valid.
4. Other criteria such as an absolute potential of less than −720 mV may be more applicable than the polarisation/depolarisation criterion in cases of immersed or underground reinforced concrete where there is limited oxygen access and polarisation and depolarisation are too slow for reliable measurement.
5. There are other constraints on the settings of the system such as minimising the risk of hydrogen embrittlement and avoiding pitting of bare titanium wires on some anode systems.
6. Modern digital systems allow convenient and accurate recording and transmission of monitoring data allowing the engineer to assess the system with minimal time on site and with far more data than simple manually controlled systems.

REFERENCES

1. Ansuini, F. J. and Dimond, J. R. Long-term field tests of reference electrodes for concrete—ten year results. NACE Corrosion 2001. 2001 Mar; Paper No. 1296.
2. Bennett, J. E. and Broomfield, J. P. Analysis of studies on cathodic protection criteria for steel in concrete. *Mater. Perform.* 1997 Dec; 36(12):16–21.
3. BS EN ISO 12696 (2012) *Cathodic protection of steel in concrete*, British Standards Institute, London, UK.
4. Stern, M. and Geary A. L. Electrochemical polarization I A theoretical analysis of the shape of polarization curves. *J. Electrochem. Soc.* 1957 Jan; 104(1):57–63.
5. Bennett, J. E. and Broomfield, J. P. Analysis of studies on cathodic protection criteria for steel in concrete. *Mater. Performance.* 1997 Dec; 36(12):16–21.
6. Glass, G. G., Hassanein, A. M., and Buenfeld, N. R. CP criteria for reinforced concrete in marine exposure zones. *J. Mater. Civ. Eng.* 2000 May:164–171.
7. Bennett, J. E. and Bartholomew, J. J. Cathodic Protection of Concrete Bridges: A Manual of Practice. Strategic Highway Research Program Report. 1993; SHRP-S-372.
8. Daily, S. F. and Green, W. Galvanic cathodic protection of reinforced and prestressed concrete using CORROSPRAY™ a thermally sprayed aluminium alloy. Corrpro technical Library paper CP-51. 2007.

Chapter 12

Case studies of cathodic protection installations

*Hernâni Esteves, Rene Brueckner, Chris Atkins,
Tony Gerrard, and Ulrich Hammer*

CONTENTS

12.1 CASE STUDY 1: AN INNOVATIVE REPAIR AND REFURBISHMENT OF AN UNDERGROUND CAR PARK IN DRESDEN, GERMANY

12.1.1 Introduction

In 1998, near to the Frauenkirche in the city centre of Dresden (Figure 12.1), the underground car park was built. The two storey underground car park can accommodate approximately 150 vehicles on an area of 4200 m² in total. The underground car park (Figure 12.2) is adjacent to the historic structure "Albertinum" and right above is the main feeder route to the Frauenkirche and the adjacent Hotel the Salt Lane.

Figure 12.1 Frauenkirche in the centre of Dresden illuminated at night.

Figure 12.2 Construction of the car park in 1998.

Damage to the garage was first observed in 2011. On the floor slab in the second basement, numerous cracks were visible, the existing coating had flakings, the underlying epoxy resin mortar was splashing and the chloride levels were elevated (Figure 12.3).

The basement ceiling in the first basement showed just as several cracks (Figure 12.4), but with higher chloride levels along with first signs of chloride-induced corrosion cracking and weeping in perimeters range, such as walls.

Furthermore, moist cracks, paint peeling and chlorides were present in the ceiling above the first basement.

Figure 12.3 Cracks on the floor slabs in the car park.

Figure 12.4 Damages and signs of moisture on the ceiling above first basement floor.

To determine the state of the topside of the ceiling above the first basement, soil pits were created in the Salt Lane. It was found that there was no seal on the ceiling. With the aid of preliminary investigations, a careful determination of costs and feasibility to approach rehabilitation of the garage was set (Figure 12.5).

One approach was to seal the topside with a waterproofing membrane. In coordination with urban issues and the adjacent Albertinum, it was realised that a seal according to the rules of the art would be possible but would mean work with highest complexity, associated with a very long repair period and resulting in high costs could be possible. A major challenge would have been to tackle the numerous cablings and urban water supply on top of the garage ceiling such as drinking water, telecom, gas, electricity lines and road drainage.

Furthermore, there were many unknown facts like the current connection situation of the garage ceiling to the existing retaining wall to the Albertinum.

Another approach was the idea to repair the garage applying the protection principle "K" according to the German Committee for Reinforced Concrete. The repair principle K means "cathodic protection" (CP).

12.1.2 Cathodic protection

CP means to trigger a current flux between the reinforcement and an additionally installed anode to gain a cathodic polarisation of the reinforcement. One method is the impressed current cathodic protection (ICCP) with an external power unit between anode and steel. Alternatively, sacrificial systems use ignoble anode materials to polarise the steel.

Figure 12.5 Investigation of the ceiling from the top.

12.1.2.1 Protective effects

The major effect is the polarisation of the reinforcement because of the impressed protection current. A sufficient current density significantly reduces or even stops corrosion completely. The protective effects may be subdivided into two categories:

- Primary effects:
  - Electron overflow in the reinforcement directly hinders the anodic dissolution, pushing the equilibrium of the reaction into the cathodic reaction direction (oxygen reduction).
  - Lowering the steel potential extremely reduces the likability of forming pitting corrosion in presence of chlorides.
- Secondary effects:
  - Formation of hydroxyl ions leads to increasing pH level at reinforcement surface.
  - In the long run, reduction of chloride content at reinforcement surface because of migration.

12.1.2.2 Advantages

- Chloride-contaminated/carbonated concrete does not need to be removed and corroding reinforcement does not need to be uncovered
- Chlorides inside the concrete matrix move away from the reinforcement
- Realkalisation of the concrete in the vicinity of the steel
- Integrated control systems give information about the structure's condition at any time
- Repair works can mostly be executed when the object is reducing disruption
- No formation of macro-cells

12.1.2.3 Cathodic protection design and layout

- BS EN ISO 12696: Targets, protection criteria and lifetime of a CP system must be clearly defined in the contract.
- A CP system may never cause or accelerate reactions which jeopardise the structures safety and use.
- The CP system needs to be adjusted to the structure, its properties and the environment.
- The layout must ensure a homogenous current distribution.
- Definition of which part of the reinforcement needs to be protected, calculation of steel surface, selection of anode type (Figure 12.6).

Figure 12.6 Layout of the cathodic protection system for the ceiling based on discrete anodes.

- Type, number and position of the monitoring sensors should be chosen in a way that the protective effect can be displayed with sufficient accuracy at any time.

12.1.2.4 Cathodic protection pre-investigations

- Continuity of the reinforcement
- Potential mapping
- Concrete resistance measurements
- Concrete cover measurements

12.1.2.5 Cathodic protection layout

- Definition of protection zones
- Making of execution plans
- Voltage drop calculation of anode system and global cabling
- Set up a quality handbook according to BS EN ISO 12696 protection criteria
- Set up a service book

Here, specifically for the garage ceiling, a system with titanium rod anodes (durAnodes3) with different lengths 150 mm up to 700 mm was selected and a special inactive length was developed to prevent current spread to the lower rebar layer (Figure 12.7).

Cracks of the ceiling were permanently sealed by injection. The concept therefore constituted a protection of the reinforcement and sealing from the inside of the garage, since without a sealing the ongoing penetration of water and chlorides from the top would make it impossible to prevent reinforcement corrosion. All cracks had to be closed before.

Along with this measure came the entire repair of the garage. Besides the ceiling above the first basement, both the floor in the first and second basement were treated with CP. Clear advantages in this project by applying CP were the fast construction time, the avoidance of deep interventions in the building by

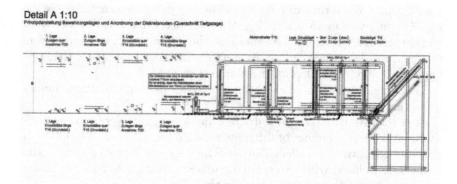

Figure 12.7 Installation of discrete anodes in hammer-drilled holes in the ceiling and adjacent bored pile walls.

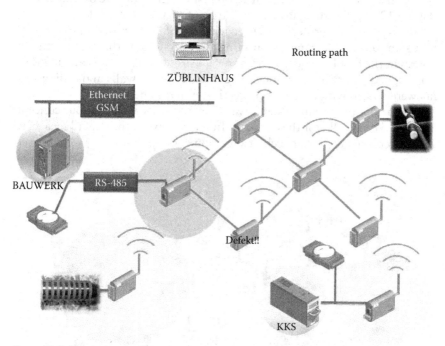

Figure 12.8 Routing path of the monitoring system.

ablation or demolition and reconstruction and the chance to actively monitor the condition of the reinforcement and to adjust the system to individual needs (Figure 12.8).

As a result of this extensive monitoring, repair intervals are extended up to 50 years.

In a construction time of only 5 months, the garage was repaired by the Ed. Züblin AG, Division Building Maintenance. At the end, more than 13,000 discrete anodes durAnodes3 of the company CPI Limited were mounted on the garage ceiling, each anode installed into a hole made by hammer drilling up to depths of 750 mm (Figure 12.9).

Each hole was individual checked prior to installation of discrete anodes to prevent possible shorts.

After the installation of these discrete anodes, anodes were connected by titanium wire to a number of different anode areas. More than 8,000 m of titanium wires were installed. The titanium wires were applied in slots and immediately closed with cover. Furthermore, sensors, reference electrodes and cathode connections were made in the ceiling.

The floor areas are protected by titanium mesh. The titanium mesh was applied to a total area of 3,900 m² on the two basement levels. After the installation of the leads, reference electrodes and cathode terminals, the entire surfaces were covered with a polymer cement concrete overlay (PCC) (Figure 12.10). For this purpose, 421 tons of PCC I were needed.

Finally, a surface protection system called OS8 was applied onto the bottom slab and to the ramp connecting the second and the first basement floor. On the first basement floor, a surface protection system called OS11a was applied. All rising components such as columns, walls and ceiling soffits were coated with a surface protection system called OS4.

The surface protection systems were in accordance with a colour concept developed especially for the car park. In the context of the overall measure, Züblin replaced the entire technical equipment.

Figure 12.9 Anode installation on the first basement level.

Figure 12.10 Casting of the PCC onto the floors to embed the anode mesh.

Figure 12.11 Entrance of the car park after the repair works were completed.

Thus, the sprinkler system, fire alarm system, the video surveillance, the entire electrical system, lighting, air conditioning and the fire alarm system had been modernised.

The electric lines, the warning systems and the supply lines for the cathodic corrosion protection system were placed in trunking in the soffit. This trunking also houses the lumineres for illumination of the garage (Figures 12.11 through 12.13).

Figure 12.12 Parking area with new colour concept after completion and media channel after conclusion.

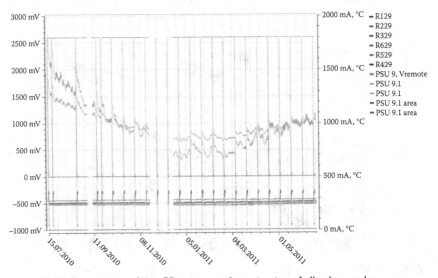

Figure 12.13 Supervision of the CP system and monitoring of all relevant data.

12.1.3 Conclusion

The CP system was commissioned in April 2010 and since that time all protection areas are fulfilling the DIN EN 12696 protection criteria.

Based on good cooperation with the client, the local team and the expert planners, the project was completed successfully and in time.

12.1.4 Case Study 1 Literature

BS EN ISO 12696:2012: *Cathodic Protection of Steel in Concrete*. London UK: British standards institute.

Chess, P., Gronvold and Karnov (1998). Cathodic Protection International, Copenhagen, Denmark. *Cathodic Protection of Steel in Concrete*. London: E & FN Spon.

Frank Muhsau, Ingenieurgesellschaft der Bauwerkserhaltung mbH, BE_BP055-01_Untersuchungsbericht.pdf, TG Coselpalais Dresden, 2009.

Bazzoni, A., Bazzoni, L., Lazzari, L., Bertolini, L., and Pedeferri, P. (1996). Field application of cathodic prevention on reinforced concrete structures. Corrosion/96, paper 312. Houston, TX: NACE.

Broomfield, J. (2007). *Corrosion of Steel in concrete, understanding, investigation and repair*. London: E& FN Spon.

Broomfield, J. L. (1987). *Cathodic protection for reinforced concrete—its application to buildings and marine structures*. Houston, TX: National Association of Corrosion Engineers.

Stratfull R.F. (1959). Progress report on inhibiting the corrosion of steel in a reinforced concrete bridge. *Corrosion* 15(6): 331t–334t.

Force Technology. http://www.forcetechnology.com. http://www. forcetechnology .com/NR/rdonlyres/BC0082BA-2F2B-4D19-B1C52D72E8E5ED8A/0/110322_CorroWatch_web.pdf.

Sensortec GmbH. http://www.Sensortec.de, http://www.sensortec.de/sensorensensors/multiring-elektrode-multiring-electrode.

12.2 CASE STUDY 2: SILVER JUBILEE BRIDGE

12.2.1 Introduction

A range of corrosion prevention techniques have been applied on the near Liverpool, England since 1993, Lambert and Atkins (2007). The techniques used range from holding repairs with corrosion inhibitors over different CP systems to the application of an electro-osmosis system.

The Silver Jubilee Bridge was constructed in the 1960s and is a Grade II listed structure (Figure 12.14). It is part of a major regional highway route that carries over 90,000 vehicles per day on four lanes over the River Mersey between Runcorn and Widnes. Any closure would result in a diversion of at least 40 mi. (65 km). Even partial closure results in heavy congestion, requiring the need for night-time maintenance adjacent to live traffic, so it is crucial to maintain the integrity and durability of the structure.

The central span of the bridge is a 330 m long steel arch structure with two 76 m side spans and is believed to be the largest of its type in Europe. The deck is reinforced concrete supported on structural steelwork. The approach viaducts have four main beams supported by reinforced concrete piers. The ends of the beams were precast, and the central spans were cast in place at the same time as the deck. The approach spans are a total of 522 m in length.

Figure 12.14 Silver Jubilee Bridge.

12.2.2 Cause of corrosion

The highways in this part of England are subjected to chloride-based de-icing salts during the winter months. The original waterproofing system of the bridge deck could not prevent extensive chloride contamination and degradation. The viaducts have joints over every third pier that have degraded with time, allowing chloride-contaminated water to leak onto the substructure. Chlorides have penetrated the concrete cover, and levels at the reinforcement have reached more than 2% by mass of cement—more than sufficient to initiate and sustain corrosion. Extensive concrete delamination could be observed on every third pier (see Figure 12.15).

12.2.3 Repair history

Several repair strategies have been used at the Silver Jubilee Bridge, such as patch repairs, electro-osmosis protection and CP systems. Most of the areas protected have been accessible from underneath the bridge, which means that although access has sometimes been extensive, it has been relatively straightforward. Access to the bridge deck was limited as it is 40 m above the River Mersey and the adjacent Manchester Ship Canal.

12.2.3.1 Holding repairs

The holding repairs carried out were mainly performed to ensure public safety. Although reinforcement section loss was not significant enough to warrant structural concerns, the public was at risk from falling delaminated

Figure 12.15 Deteriorated approach viaduct pier.

concrete from under the approach viaducts. Loose and delaminated concrete was removed and the steel and exposed concrete overcoated using a polymer-modified cementitious mortar containing an amino alcohol corrosion inhibitor. The mortar coating minimised any further corrosion of the reinforcement and prevented significant further ingress of the contaminants. Concrete was not reinstated to prevent incipient anode effects where the repaired areas become cathodic to the adjacent areas causing enhanced corrosion of the surrounding steel.

This repair method has been performing adequately and prevented significant section loss over the previous 10 years, Baldwin and King (2003). Minor issues such as discoloration of repair areas, which is an anticipated side effect due to ultraviolet exposure, and slight degradation of the coating were observed.

12.2.3.2 Electro-Osmosis

The application of a DC current across a porous solid can generate movement of moisture because of electro-osmosis. Although techniques such as CP, electrochemical chloride extraction (ECE) and realkalisation result in a small effect, electro-osmosis as a technique in its own right has been applied to the movement of moisture through porous materials such as concrete, masonry or soil for a considerable time and with varying results.

An electro-osmosis system was specifically developed to control moisture levels in a pier of the bridge by the application of controlled low voltage DC pulses. The system is capable of reducing moisture levels in concrete to between 60% and 70% RH and maintaining this level irrespective of external weather conditions. According to Vernon (1935), corrosion of steel commences at a slow rate at approximately 60% RH and significantly increase at 75–80% RH. These thresholds may be affected by the level of chloride contamination. The pier was suffering from alkali—silica reaction, which can also be controlled by reducing the available water.

The system was also designed to negatively polarise the reinforcement resulting in a degree of CP, helping to reduce the corrosion risk of embedded steel during the transition period from high to low relative humidity (typically several months) and providing additional protection throughout the life of the installation (Lambert 1997).

The system has been specifically assessed for possible side effects resulting from its operation. No evidence has been found to indicate significant risks of bond strength reduction, excess alkali generation, hydrogen evolution or stray current corrosion of adjacent discontinuous steel. This was the first such system in the United Kingdom. It was installed in 2002 and has sufficiently protected the trial pier. The egress of excess moisture can be seen in Figure 12.16.

12.2.3.3 Cathodic protection

Reinforced concrete can be cathodically protected using various methods by means of an ICCP or a galvanic system. Both systems work by polarising the reinforcement in an electrical circuit so the anodic, iron-dissolving

Figure 12.16 Egress of excess moisture.

mechanism is forced to take place at an artificially generated anode. ICCP systems generally use inert long-life electrodes such as mixed metal oxide— coated titanium. The reinforcement is polarised using an external DC power source. Galvanic systems use less noble metal electrodes, commonly zinc, aluminium or magnesium, which corrode preferentially to the steel and thereby provide the required protection.

12.2.3.3.1 Conservative cathodic protection systems

The first CP systems at the Silver Jubilee Bridge were installed at two piers in 1993 comprising each 5 zones, 25 Ag/AgCl KCl and 18 graphite pseudo- reference electrodes. These CP systems represented the first major use of coated titanium mesh with a dry-mix sprayed concrete or gunite overlay in England. The basic mesh and overlay system was also used on the next two repair schemes, but in 1998, a combination of discrete anodes and mesh and overlay was used. Dry-spray gunite was used for the concrete repairs, whereas wet-spray gunite was used for the overlay to minimise dust and noise disruption to a neighbouring school.

In 2000, a CP system and patch repairs were used during a repair con- tract, which included extensively contaminated areas next to the abutment and locally affected areas at the highest sections of the approach viaducts. To reduce costs, locally affected areas with difficult access were patch repaired using hand-applied mortar containing corrosion inhibitor to pro- tect against the incipient anode effect.

The systems installed after 2000 were redesigned by reviewing the operating criteria of the existing systems. By using this technique it was possible to reduce the quantity of anodes used, in some cases by a factor of 3. In addition, the lower current demand meant larger zones could be used, which reduces the number of monitoring probes and power supplies. As an example, one system installed in 1995 used 7 zones to protect one 30 m long beam. Four of these zones had multiple layers of anode mesh. There were 24 Ag/AgCl/KCl and 12 mixed metal oxide (MMO) coated titanium pseudo reference electrodes and 6 graphite potential probes installed. In 2005 a similar beam was protected as a single zone, with a single layer of mesh MMO Ti anode and 4 reference electrodes. All systems were designed to have at least 4 No. Ag/AgCl/KCl reference electrodes per zone independent of their area. The majority of Ag/AgCl/KCl reference electrodes installed during the first installations are still operational.

In 2011, further pier and beam elements of the approach viaducts have been protected using either discrete or mesh and overlay systems. These systems consist of 2 Ampere zones with each 4 to 6 reference electrodes. Elements within the same span were included in one single zone and a pier requiring protection was protected using one single zone with 4 reference electrodes. This is a substantial reduction to the adjacent pier which was protected in 1993 having 5 zones with 25 Ag/AgCl/KCl and 18 graphite pseudo reference electrodes.

12.2.3.3.2 Deck cassette system

There were limitations in respect of the suitability of systems to be installed on the deck. Accessibility was a major factor as the bridge deck is 40 m above the River Mersey and Manchester Ship Canal. Traffic management is restricted and any closures had to be avoided. Other factors affecting the suitability were imposed by the structure. The deck is 200 mm thick and exposed to constant vibration caused by the traffic.

A mesh and overlay system would be able to provide the required current, but the vibrations in the deck would cause a high risk of debonding of the sprayed concrete overlay once it has been installed using extensive access. Discrete anodes could be installed by roped access, but would require drilling holes into the deck at depth. If the holes were drilled marginally too deep, there was a risk of drilling into live traffic. The application of corrosion inhibitors could provide a time limited protection of approximately 10 years in areas of chloride contents less than 1% by mass of cement. However, the deck was mainly contaminated with chlorides up to 2% and a more durable solution was required.

The Protector Intranode cassette system was identified to be the most suitable system as it would avoid the only remaining repair option, the removal and replacement of the deck, Brueckner and Lambert (2011). Aside from the traffic chaos associated with its closure, the environmental consequences of this could not be tolerated. The deck contains approximately 1000 m³ of concrete. The embodied energy in the deck concrete is around 6 Terajoules, equivalent to the energy produced by 1200 barrels of oil. The diversion of the traffic over the 40 mi. long alternative route would cause the release of an additional 1000 tons CO_2 per day based on 20% heavy good vehicles (HGV) for the total 90,000 vehicles crossing the bridge per day. This does not take account of the associated disruption to the local economy and population, which has been valued at £160,000 per hour.

The cassette system has originally been used on harbour structures in Norway but there was no track record of installations on other structures. The system consists of MMO coated titanium ribbon anodes in a glass fibre filled FRP tray, which can be mounted on a concrete surface using sleeved bolts (see Figure 12.17). In the environment of jetties and harbours, the glass fibre foam is regularly wetted by the tide providing the required electrolyte. The original system was improved using calcium nitrate impregnated glass fibre foam, which is able to remain moist simply by being in contact with the atmosphere.

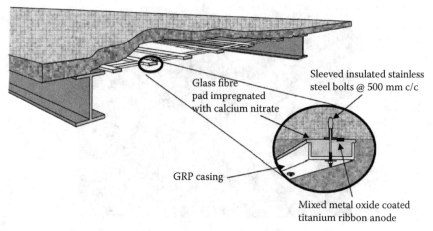

Figure 12.17 Cassette system installed on deck soffit.

As the system has never been used before on other than marine structures in Norway, its suitability was assessed on a 60 m long trial area on the soffit of the deck. The surface area to be protected was 520 m². The trial comprised Protector Intranode cassettes installed at 0.5 m centres in five zones of approximately each 100 m², which were monitored with 7 No. Ag/AgCl/KCl references electrodes. The five zones were monitored over a period of 1 year and showed satisfactory results so that the remaining deck could be protected using this system. The design was reviewed on the data obtained from the trial and the area per zones could be doubled to 200 m² with a maximum current output of 2 A. The number of references electrodes was decreased to a minimum of 4 per zone. The remaining 3320 m² of the deck were protected in 17 zones. The design was based on a current density of 10 mA/m².

12.2.3.3.3 Pre-stressed beam cathodic protection system

A galvanic CP system was installed on the soffit of 6 No. 14 m long pre-stressed beams of an approach viaduct of the bridge in 2011. The pre-stressed elements of the approach viaduct were constructed later than the piers and deck when the bridge was widened in 1977. Problems with the joints and drainage led to chloride contamination of the pre-stressed beam soffits causing delamination (see Figure 12.18). The beam soffits have been protected using zinc layer anodes as it was found to be relatively straightforward to install and safe with respect to hydrogen embrittlement of the pre-stressing wires.

In order for CP to work as with in situ reinforced concrete, it was necessary to ensure all the steel is electrically continuous. The techniques used for conventional reinforced concrete such as welding or shot fired can be

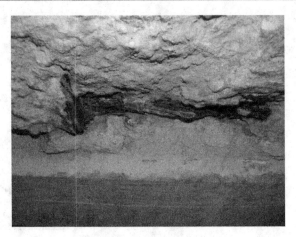

Figure 12.18 Corrosion of pre-stressing wires in precast beam.

used on the link reinforcement but was considered to be inappropriate for the pre-stressing wires. It was found to be more practical to use a metallic strap, cable tie or tie wire to make the wires and reinforcing steel continuous. It should be noted that many pre-stressed beams have large amounts of reinforcement installed in relatively small cross sectional areas and so there is a reasonable expectation that continuity may be adequate within an individual beam. However, isolated tendons were found and an accepted detail for bonding in was developed. Continuity is unlikely to be present from one beam to the next as they are separate elements but this was addressed during the installation phase.

The anode system was designed to run along the beams. This removed the need to consider the potential for relative movement between beams and the anode's ability to tolerate this. The breakouts had to be carried out with care because of the residual stress in the elements and the structural consequences of the loss of bond in the tendons. It was important to control the extent of the breakout in length, width and depth, which was calculated by a structural engineer. Hydrodemolition proved to be an appropriate method to expose tendons on the beams.

The exposed surface of pre-stressed concrete is generally limited to the soffit of the beams. This is likely to be the area where corrosion takes place because of the availability of oxygen. It is also likely that this would have the highest level of chlorides present due to the evaporation of free water from the soffit, which was confirmed on the beams.

Regular monitoring will provide data about the performance of the systems and most likely lead to the protection of other pre-stressed bridge elements in the near future.

To monitor impressed current systems, they are designed to be switched off, and the potential decay curve checked. This is straightforward as they have to be designed so that the anode and cathode are electrically isolated. For a galvanic system, this isolation is not necessary and so this provides an opportunity to simplify installation. Short circuits are, in effect, a benefit but prevent depolarisations being undertaken. For this system, it was decided to install permanent reference electrodes into the concrete and collect data for 7 days before any galvanic anodes being installed. The potential shift achieved when the anodes installed was considered indicative of protection being achieved. It will also be used to identify when the anodes are no longer providing protection. IR error (see Section 11.2.3.1) is not considered a significant issue since the reference electrodes are installed close to the steel and the current passed from a galvanic anode is low, and so the IR error will be low.

12.2.4 Rationalisation of cathodic protection systems

The existing approach viaducts and deck of the structure have had numerous CP systems installed dating back to 1993. Electronic equipment and software control have undergone huge developments during the past two decades. Also different control systems have been used. This resulted in difficulties maintaining the various systems. Additionally many systems are out of date and require a separate phone line, which needs to be maintained.

After completing the protection of the deck, it was considered to review all systems based on the data and experience gained during monitoring. The control and monitoring equipment was rationalised based on the following criteria:

- Existing control systems to be replaced
- Reduction of number of zones and probes to be monitored
- Direct monitoring and adjustment via physical site connection or keypad on board of system.
- Remote monitoring and adjustment via a dial up. Maximum 3 phone lines for the two approach viaducts and deck
- Reduction in number of zones, while maintaining ability to adjust sub zones.
- Ability to monitor a minimum of 4 reference electrodes per zone
- Direct measurement of current and DVM.
- Collection of monthly instant off data (or on demand).
- Collection of monthly 24 hour depolarisation data (or on demand).

The review of the data of the systems installed until 2005 showed that it was possible to significantly reduce the number of zones and therefore also the amount of reference electrodes to be monitored per system. For example, two systems with a total of 15 zones were combined into one system with only 2 zones. Five to six zones were reduced to 1 or 2 zones with 4 reference electrodes per zone to be remotely or physically monitored using the main control units. The redundant reference electrodes are still available to be monitored in a sub-control unit level, if any future failures should occur.

The deck CP system is not affected by the rationalisation, despite the increased amount of zones and reference electrodes of the trial section compared to the main deck CP installation. This CP system is still relatively recent and therefore an increased amount of monitoring data is considered to be beneficial to design protection system for other structures in the future.

CASE STUDY 2 REFERENCES

Baldwin, N.J.R. and King, E.S. 2003. Field Studies of the Effectiveness of Concrete Repairs, Phase 4 Report: Analysis of the Effectiveness of Concrete Repairs and Project Findings. Research Report 186, Health and Safety Executive, Sudbury, UK.

Brueckner, R. and Lambert, P. 2011. Renovation of the Deck of a Major Listed Bridge Structure. STREMAH 2011—Structural Studies, Repairs and Maintenance of Heritage Architecture XII, Tuscany, Italy, WIT Press, Southampton, pp. 307–317.

Lambert, P. 1997. Controlling Moisture, Construction Repair: Concrete Repairs 6.

Lambert, P. and Atkins, C.P. May 2007. Maintaining the Silver Jubilee Bridge, Concrete International, ACI, pp. 53–57.

Vernon, W.H.J. 1935. A Laboratory Study of the Atmospheric Corrosion of Metals. Trans. Faraday Soc. 31, pp. 1668–1700.

12.3 CASE STUDY 3: GALVANIC ANODE SYSTEMS

12.3.1 Aluminium-indium-zinc metal sprayed galvanic anode

When first constructed in 1998–1999, one small section of the northbound carriageway wall of a motorway bridge was found to have used mixing water from a borehole with very slightly raised chloride ion concentration levels.

It was decided that a targeted ICCP system based on proprietary impressed current discrete anodes would be the most appropriate. The CP system was installed within a year of the culvert being constructed so very little time passed to allow corrosion to initiate.

The walls are almost 1 m thick and have good concrete cover depths to protect the steel from corrosion.

Because of continued vandalism, an alternative replacement CP system needed to be designed.

The replacement cathodic prevention system was based on a sprayed proprietary galvanic anode of aluminium–zinc–indium alloy applied directly to the surface of the contaminated wall. All direct current (DC) and monitoring fixings will be embedded into the concrete so that no components are on show to attract vandalism or appear to have any value.

The surface of the concrete receiving the replacement cathodic prevention system was prepared by abrasive grit blasting to remove surface laitance and clean the surface. The proprietary anode metal spray was then applied to the prepared surface.

Individual cathodic prevention components such as DC and monitoring connections and reference electrodes were all buried in the concrete, again to minimise the risk of vandalism. The connections between the steel reinforcement and the anode are proprietary and were made especially for this project.

12.3.2 Surface mounted zinc sheet galvanic anode

The soffits of a number of pre-stressed beams on an approach viaduct built in the mid 1970s and adjacent to a tidal estuary were showing signs of corrosion and deterioration due to chloride ingress both from the environment and surface water leakage from the road above.

Both the location and the need to keep protective potentials to values that would not cause problems to the pre-stressing elements of the beams a galvanic anode system utilising zinc sheet was specified and installed.

Each beam was mechanically cleaned to ensure a debris free surface and continuous zinc strips 0.25 m wide and 250 µm thick installed along the centre line of the soffits.

The zinc has a conductive ionic gel on the rear face which allows electrical conductivity between the zinc and reinforcement steel through the concrete, as well as providing some adhesion to the concrete surface.

The tape was additionally held in place by expanding concrete anchors. Both for aesthetic purposes and to ensure that moisture could not ingress along exposed sides of the hydroscopic gel the whole surface of the Zinc and for 20 mm onto the concrete was covered by a waterproof cementitious screed.

Reference electrodes and make or break connections for monitoring purposes were installed at various locations.

Chapter 13

Economic aspects

Paul Lambert

CONTENTS

13.1 INTRODUCTION

The effectiveness of cathodic protection for the full- and long-term repair and life extension of corrosion-damaged reinforced concrete is thoroughly established and recognised by many national and international bodies including the American Concrete Institute, U.S. Federal Highways Authority and the U.K. Highways Agency. For the technique to maintain and further develop this position, it must be seen to be both technically and economically attractive. It is also now increasingly important for any such technique to be environmentally friendly and justifiable on a sustainability basis.

Cathodic protection can be used as a part of the repair strategy virtually at any instance of reinforcement corrosion including that resulting from carbonation. It is in the area of chloride contamination where the cost advantages have historically been most apparent as the technique avoids the need to remove sound- but chloride-contaminated concrete while still providing an essentially permanent repair solution.

The ability to leave structurally sound-contaminated material in place has not only advantages in terms of cost but also benefits sustainability, health and safety while minimising disruption to the general public. As a result, the technique has been widely adopted for the life extension of reinforced concrete bridge structures, most notably in the United States and Europe (Transportation Research Board, 2009).

Although traditional cathodic protection on pipelines, ships, rigs and jetties tends to be predominantly applied from new, with reinforced concrete, it has most commonly been employed as part of a repair strategy for structures already suffering from or at risk of serious reinforcement corrosion. The number of applications to new structures (so-called cathodic prevention) still remains relatively small; yet, it is this area where the greatest economic and sustainability benefits may be obtained.

Properly designed and applied, an integral system operating from new demonstrates that prevention is not only better than cure but significantly cheaper and less disruptive to the structure, the general public and the environment.

13.2 GENERAL COST IMPLICATIONS OF REPAIR

Conventional repair of chloride-contaminated reinforced concrete can be a very expensive and time-consuming exercise. The U.K. Department of Transport estimated in 1989 that more than £500 million was needed to be spent on the cathodic protection of motorway and trunk road bridges alone, which constitutes only 10% of the total U.K. bridge stock (Wallbank, 1989).

Around the same period, the United States with over 200,000 bridges suitable for cathodic protection was estimated to require more than $20 billion to complete the work, a potential saving of around 75% on replacement (Wyatt, 1993). Soon after, a U.S. Federal Highways study identified that around 15% of U.S. bridges were structurally deficient due to corrosion. At that time, the annual direct cost for repair and maintenance of highway bridges was estimated at around $8 billion and that it would cost up to $100 billion to upgrade, repair or replace the existing backlog (Federal Highways Administration, 2001).

Following the I-35 bridge collapse in Minneapolis in 2007, the Federal Highways Authority National Bridge Inventory showed over 25% of bridges to be deficient and in need of repair or replacement, with the figure being over 40% in many key states (Larsen, 2008). Further studies of the U.S. bridge condition data have helped define the nature of the problems and the potential benefits of techniques such as cathodic protection in their remediation (Lee, 2012a and b).

In the United Kingdom, as elsewhere in Europe, it has been suggested that the equivalent of more than half the annual construction budget is now spent on the repair of existing structures (Waterman, 2006). Figures from the U.K. Office for National Statistics for 2011 appear to confirm this with around £15 billion being spent on new infrastructure and £8 billion on infrastructure repair during the period.

13.3 CONVENTIONAL REPAIR VERSUS CATHODIC PROTECTION

When patch-repairing chloride-contaminated structures, it is necessary to remove large quantities of material to ensure that corrosion does not reinitiate. By avoiding the need to remove contaminated material, cathodic protection can offer significant financial benefits when compared with more conventional repair strategies. Additionally, cathodic protection will prevent future corrosion, even where the source of contamination cannot be removed. In this way, not only is the initial investment potentially smaller but the need for costly future intervention is greatly reduced.

The choice of remedial techniques to be applied to a contaminated or corrosion-damaged reinforced concrete structure will have a significant influence on a number of costs associated with the repair. These can generally be divided into three areas: costs directly associated with the repair technique; indirect costs necessitated by the choice of repair technique and ongoing maintenance costs following repair.

13.3.1 Direct costs

Direct costs should be relatively easy to estimate and can be readily compared for various repair options. For conventional repairs of chloride-contaminated concrete, direct costs include the following:

1. Identification of chloride-contaminated concrete
2. Removal of chloride-contaminated concrete
3. Surface preparation of reinforcement
4. Replacement of badly corroded reinforcement
5. Reinstatement of the concrete

For cathodic protection, direct costs include the following:

1. Initial system design
2. Breakout of cracked or spalled concrete only
3. Surface preparation of exposed reinforcement
4. Replacement of badly corroded reinforcement
5. Continuity testing
6. Bonding of discontinuous reinforcement
7. Reinstatement of concrete
8. Installation of anode and monitoring/control system

It can appear that there is significantly more effort and therefore cost associated with the installation of a cathodic protection system. However, this is generally not the case. In instances where isolated chloride contamination is present over relatively small areas, it is usually the case that conventional local breakout and reinstatement to the contaminated areas are the most suitable routes to a successful and cost-effective repair.

Structures that have undergone a very high degree of deterioration and that require the removal of large volumes of concrete regardless of the repair technique adopted may show little or no cost advantage from employing cathodic protection rather than more conventional means. In such cases, it is usually prudent to undertake a cost-benefit analysis to determine the most economic long-term repair strategy. This should help to identify the relative importance of factors such as the cause of deterioration and the likelihood of preventing further contamination. Such factors can adversely affect the durability of conventional repair but are not always immediately obvious.

The greatest economic benefit from a cathodic protection is where repairs are required for structures that contain high levels of chloride contamination but that have yet to undergo extensive deterioration through reinforcement corrosion and delamination. In these cases, the fact that large quantities of contaminated concrete can remain in place by adopting cathodic protection as the principal repair method can dramatically reduce costs. The removal and disposal of chloride-contaminated concrete and reinstatement, for example, with a good quality sprayed concrete repair material, remains a high cost activity and therefore every cubic metre saved from replacement constitutes a considerable saving.

13.3.2 Indirect costs

It is often with regard to indirect costs that cathodic protection shows benefits over conventional patch repair. If one considers a contaminated reinforced concrete bridge structure suffering from serious chloride contamination, then conventional repair of such a structure may require the design, approval, installation, maintenance and removal of a temporary support structure. This can be an expensive and a time-consuming exercise and the presence of such temporary supports, no matter how well designed, may require load and speed restrictions to be applied to the traffic on the carriageway above.

Where complete demolition and replacement are required, the level of intrusive traffic management associated with such works may be so disruptive as to make it unacceptable except where absolutely unavoidable. The use of cathodic protection can dramatically reduce the need for such measures.

Buildings incorporating reinforced concrete structural members or pre-cast units such as cladding panels are further potential candidates for the benefits of cathodic protection. Now largely forbidden or controlled in most countries, chloride-based admixtures were widely used as set accel-erators until comparatively recently. While allowing faster, more economic production rates, especially in cold weather, the addition of chloride to fresh concrete can have serious consequences with regard to the durability of the structure.

Once constructed, the cost of major repair or replacement of such ele-ments with the inevitable disruption to tenants and other users of the build-ing is often prohibitive in all but the most exceptional cases. Under these circumstances, cathodic protection offers an economic alternative with minimal disruption to the tenants, particularly in non-domestic structures where work can be programmed for nights and weekends.

13.3.3 Maintenance costs

It is tempting to assume that once a conventional repair has been properly undertaken, there will be little or no future maintenance costs and that the structure is somehow better than new. This is rarely the case except in the most exceptional circumstances. Even where potential sources of future contamination have been dealt with, such as through the replacement of leaking movement joints or the application of a protective coating system, a continuing level of maintenance will be required periodically during the full life of the structure. Where the source of contamination has not been properly addressed, the whole sequence of remedial works can start again within a very short timescale.

Cathodic protection is no different in requiring a commitment to long-term maintenance, but in general, the costs are relatively small and can be readily identified and quantified. The costs associated with the maintenance of a cathodic protection system fall into two areas. The first is the provision of the electrical current used by the system to protect the steel reinforce-ment. Typically, protection currents are very low (between 5 and 20 mA/m^2 of reinforcement), which means that many systems consume about as much power as a domestic light bulb. Although the cost of the electricity is virtu-ally negligible, there may be ongoing costs associated with maintaining the security of supply, for example, repair of damage to cable runs.

The second ongoing source of expense is associated with the monitoring and control of the system. For a manual system, this will involve periodic site visits to carry out monitoring and adjust controls. It is rare for modern systems to be manually monitored and adjusted and most now have some form of data logging or remote operation through modem, dedicated radio or internet link. Innovation and increased competition have improved the reliability and reduced the cost of such systems. Over a period of time, remote monitoring and control system can save significant expenditure, especially if the system is difficult to access due to location or operating restrictions. For a typical four-zone cathodic protection system, the UK Highways Agency quotes figures of £1500–£2500 per annum for manual assessment and reporting, which drops to £1000–£1500 per annum for remote monitoring with one visual inspection (Design Manual for Roads and Bridges, 2002).

While there are considerable benefits in employing remote monitoring and control systems, it is important to ensure that a minimum level of hands-on site inspection is maintained to confirm the proper operation of the system and to help identify potential problems at an early stage.

13.4 COST IMPLICATIONS OF ANODE TYPE

The anode selected for a cathodic protection system has implications for both the direct costs and possible future maintenance associated with the system. Numerous types and variants of impressed current anode are available, but these generally fall into one of the four generic categories as shown in Table 13.1. There has been an increased use of galvanic anode systems, generally based on zinc. These typically fall within the same price range as the impressed systems and can be used to provide up to 15 years of service with relatively little intervention although such systems are often employed to provide a degree of corrosion protection rather than full cathodic protection in accordance with the relevant international standards (Broomfield, 2002).

In the majority of cases, the selection of an anode type is based on design factors such as protection current requirements, design life, additional weight constraints and familiarity with a particular material. The direct cost of buying and installing the anode system is often secondary, particularly when considered as a proportion of the overall cost of the remedial work. There are also differences between the various systems with regard to resistance to damage and ease of repair. Such factors will have particular importance where a structure is considered to be at high risk of accidental damage, vandalism or theft.

13.5 COST COMPARISONS

An early comparison of cathodic protection repair costs was undertaken for two reinforced concrete support piers based on full replacement, conventional repair or cathodic protection. The contract involved remedial works to chloride-contaminated concrete, bearing replacement, road deck waterproofing and resurfacing to a value of approximately £500,000

Table 13.1 Performance characteristics and budget costs for various anode systems (2002 prices)

Property	Coated titanium mesh and cementitious overlay	Conductive coating	Discrete anode	Thermal sprayed zinc
	System			
Installation	Fixing of mesh can be time consuming Roughened substrate required for overlay	Surface preparation of concrete must be correct and coating applied in good weather conditions	Care must be taken to avoid hitting steel when drilling holes to accept anodes	Correct surface preparation to ensure bond to substrat May require use of a humectant
Anode service life[a]	Up to 120 years	Up to 15 years	Up to 50 years	Up to 25 years
Ease of repair or replacement	Difficult as both overlay and mesh have to be removed and reinstated	Familiar technology Re-coating is generally relatively easy	Anode can be drilled out and replaced in the event of a defect	Zinc coating can be reapplied over existing sound substrate
Budget cost[b]	£60–£100/m²	£20–£40/m²	£40–£100/m²	£60–£100/m²

[a] Service life based on average current output. Life will be reduced at unduly high currents or for inadequately designed, installed or maintained systems. Replacement systems may be able to use some of the original wiring and connections.

[b] Figures do not include for concrete repairs prior to installation of system (Design Manual for Roads and Bridges, 2002).

(Lambert et al., 1994). Comparing costs for the project as a whole, cathodic protection was found to result in a 25% reduction in the contract price compared to full replacement. The saving for cathodic protection over conventional repair was calculated to be 50%. When a straightforward comparison of the reinforced concrete remedial works was undertaken, with other elements of the works ignored, cathodic protection offered a 40% reduction in costs compared to full replacement and a 75% reduction compared to conventional repair (see Figure 13.1).

In another study, the relative costs of cathodic protection and conventional repair to a deck support beam were found to be almost identical. However, the reduction in capacity associated with the full-replacement option would have necessitated significant traffic restrictions on the road above. Due to the critical nature of the structure in maintaining traffic flows and lack of alternative routes, such traffic restrictions were deemed unacceptable. Cathodic protection was therefore selected for the remedial works (Haywood, 1995).

Before
repair

20 years
after repair

Figure 13.1 Example of a durable and cost-effective repair employing cathodic protection. (Silver Jubilee Bridge, Cheshire, photographs courtesy of Halton Borough Council.)

A more detailed analysis based on discounted cash flow techniques for the maintenance of elevated motorway structures in the United Kingdom showed even greater cost benefits for cathodic protection when considered over a 40-year period (Unwin & Hall, 1993). When compared with replacement, conventional repair showed a 10% saving whereas cathodic protection saved 85%. Significantly, if the installation of cathodic protection was delayed for 10 years, the calculated saving against replacement would have reduced to 50%.

In the period since these studies were carried out, the relative cost of cathodic protection has reduced considerably, partly through increased competition but also as a result of optimised design and improved monitoring and control systems. By comparison, the costs of conventional patch repair have remained high, not helped by enhanced health and safety requirements and the increased costs associated with the disposal of waste material.

A more recent cost comparison looked at a dual two-lane motorway overbridge requiring concrete repairs to the two central supports. The cost for patch repairing approximately 50 m² of reinforced concrete was estimated to be £38,000. The equivalent cost for applying a conductive coating cathodic protection system was less than £13,000. However, if the costs of access and traffic management were included, the total expenditure for patch repairs became £98,000 compared with £23,000 for the cathodic protection option due to shorter access times and reduced traffic management (Arya & Vassie, 2005). This once again highlights the considerable savings in expenditure that can be made through the adoption of electrochemical remediation techniques.

A recent study in the Netherlands of service life and life cycle costs for more than 100 structures with cathodic protection confirms that once operational, the systems have been effective in preventing further corrosion. This was achieved with little or no maintenance intervention for period of up to 20 years, the life of the installations at the time of the study (Polder, Leegwater, Worm, & Courage, 2012). Where intervention was required, it was minor in nature and related to poor drainage details or inadequate electrical isolation. As would be expected, the level of intervention increased with age, with around 10% needing minor maintenance at 7 years, rising to 50% for 15-year-old systems.

13.6 CATHODIC PREVENTION

The approach by which cathodic protection is applied to new reinforced concrete structures to prevent future corrosion rather than control existing corrosion is commonly referred to as cathodic prevention. Cathodic prevention has been used extensively on elevated road structures in Italy for a number of years and has been considered elsewhere for installation on bridges where corrosion-damaged elements are subject to complete reconstruction (Bazzoni et al., 1994). It is now commonly employed on reinforced concrete structures in the aggressive environment of the Middle East and on marine structures such as jetties and sea water intakes (Chaudhary, 2002).

Cathodic prevention has many cost advantages over cathodic protection. The current requirements are much lower at approximately 2–5 mA/m^2 steel as opposed to 5–20 mA/m^2 of steel for cathodic protection. The relative installation costs are also lower on new-build structures, typically 2%–3% of the cost of the works, and can lead to cost savings by avoiding the need for other protection systems such as coatings.

The presence of a properly designed, installed, monitored and controlled cathodic prevention system can ensure the prevention of future corrosion-related problems, thus dramatically reducing the long-term maintenance costs of the structure.

13.7 PROTECTING THE INVESTMENT

As with many areas of civil works, obtaining meaningful guarantees for remedial works is often impractical. Insurance-backed guarantees remain generally unavailable for cathodic protection installations on existing structures. The best way of achieving a durable repair is to ensure that all stages of the cathodic protection installation, from initial investigation to daily operation, are undertaken to the highest technical standards under rigorous quality assurance procedures and in accordance with current standards (e.g. BS EN ISO 12696, 2012). There are a number of national and international organisations and certification schemes that help maintain the level of technical competence of those involved in this technology.

Regardless of the increasing number of successful installations being reported, cost will always dominate the repair and protection of reinforced concrete and it is the proven cost-effectiveness of the technique that remains the most powerful argument for employing cathodic protection.

13.8 SUSTAINABILITY CONSIDERATIONS

In addition to the financial implications of repair options, it is becoming increasingly important to take account of the environmental impact associated with any particular system. By considering the embodied energy component of various repair options, it is possible to establish the associated carbon dioxide emissions resulting from their use (Atkins, Buckley, & Lambert, 2006).

The energy implications of different forms of repair can be compared by considering a typical representative element with the embodied energy expressed in gigajoules (GJ). For a simple panel of reinforced concrete, nominally 1 m^2 by 200 mm deep, the embedded energy of the original construction would be approximately 1.2 GJ for the concrete plus 0.8 GJ for the reinforcing steel, resulting in the generation of up to 200 kg of carbon dioxide.

Patch repairing the panel after 20 years of exposure to de-icing salts could involve breaking out chloride-contaminated material to a depth of 50 mm and reinstating with a polymer-modified cementitious repair mortar at an additional energy cost of 5 GJ, roughly equivalent to 1 barrel of oil or half-a-tonne of CO_2.

As an alternative to breakout and reinstatement, the panel could have a mixed metal oxide–coated titanium mesh and sprayed overlay cathodic protection system applied at an energy cost of around 1.7 GJ/m^2. Once commissioned, the energy consumed to operate the system should be no more than 0.004 GJ/m^2/year. Cathodic prevention represents an even greater energy saving as it requires fewer components and less power to operate. Such systems may be as low as 0.04 GJ/m^2 installed with an additional 0.002 GJ/m^2/year of operation. In this way, cathodic protection can represent a considerable potential saving in carbon dioxide generation, in addition to greatly reducing the amount of conventional repair required.

Cathodic protection as applied to reinforced concrete therefore proves itself to not only be a technically elegant and commercially favourable option for remediation and durability enhancement but also to be compatible with modern expectations of low environmental impact and sustainability.

REFERENCES

Arya, C. & Vassie, P. (2005). Assessing the Sustainability of Methods of Repairing Concrete Bridges Subjected to Reinforcement Corrosion. *International Journal of Materials and Product Technology*, 23 (3/4), 187–218.

Atkins, C., Buckley, L., & Lambert, P. (2006). Sustainability and Repair. Concrete Communication 2006, Concrete Society, UK.

Bazzoni, A., Bazzoni, B., Lazzari, L. et al. (1996). *Field Application of Cathodic Prevention on Reinforced Concrete Structures*. NACE Corrosion '96, Denver, USA.

Broomfield, J. (2002). The Principles and Practice of Galvanic Cathodic Protection for Reinforced Concrete Structures. Corrosion Prevention Association, Technical Note No. 6.

BS EN ISO 12696. (2012). *Cathodic Protection of Steel in Concrete*. British Standards Institution, UK.

Chaudhary, Z. (2002). NACE Corrosion '02, Denver, Paper 2260.

Design Manual for Roads and Bridges (DMRB). (2002). Cathodic Protection for Use in Highway Structures, BA 83/02, Highways Agency, UK.

Federal Highways Administration. (2001). Corrosion Costs and Preventative Strategies in the United States, FHWA-RD-01-156.

Haywood, D. (1995). Approach Shot. *New Civil Engineer*, 25 March, 26–7.

Lambert, P., Shields, M., Wyatt, B. et al. (1994). Runcorn Approach Viaduct: A Case Study in Assessment and Cathodic Protection of Reinforced Concrete. *Eurocorr/UK Corrosion '94*, 3, 1–23.

Larsen, K. R. (August 2008). New Legislation Focuses on Extending the Life of Highway Bridges. *Materials Performance*, 30–35.

Lee, S. -K. (2012a). *Current State of Bridge Deterioration in the U.S. Materials Performance*, Part 1—January 2012, 62–67.

Lee, S. -K. (2012b). *Current State of Bridge Deterioration in the U.S. Materials Performance*, Part 2—February 2012, 40–45.

Polder, R. B., Leegwater, G., Worm, D., & Courage, W. (2012). Service Life and Life Cycle Cost Modelling of Cathodic Protection Systems for Concrete Structures, International Congress on Durability of Concrete, June, Trondheim, Norway.

Transportation Research Board. (2009). *Cathodic Protection for Life Extension of Existing Reinforced Concrete Bridge Elements*. NCHRP Synthesis 398, Washington DC.

Unwin, J. & Hall, R. J. (1993). Development of Maintenance Strategies for Elevated Motorway Structures. *Proceedings of Fifth International Conference on Structural Faults and Repair*, 1, 23–32.

Wallbank, E. J. (1989). *The Performance of Concrete in Bridges*. HMSO, UK.

Waterman, A. (March 2006). Costs: Concrete Repairs. *Building*, March.

Wyatt, B. S. (1993). Cutting the Cost of Corrosion of Reinforced Concrete Highway Structures. *Proceedings of Fifth International Conference on Structural Faults and Repair*, 1, 129–42.

Index

Printed in the United States
by Baker & Taylor Publisher Services